T0309160

Discrete Mathematical Structures

Discrete Mathematical Structures

Jonas Hughes

www.willfordpress.com

Published by Willford Press,
118-35 Queens Blvd., Suite 400,
Forest Hills, NY 11375, USA

ISBN: 978-1-64728-504-3

Cataloging-in-Publication Data

Discrete mathematical structures / Jonas Hughes.
p. cm.
Includes bibliographical references and index.
ISBN 978-1-64728-504-3
1. Discrete mathematics. 2. Numerical analysis. 3. Mathematics. I. Hughes, Jonas.
QA297.4 .D57 2023
511.1--dc23

For information on all Willford Press publications
visit our website at www.willfordpress.com

Willford Press

Contents

Preface

Discrete mathematics refers to the study of mathematical structures which are discrete and not continuous. Logic statements, integers and graphs are some of the objects examined in discrete mathematics. It is concerned with the investigation of countable sets, which might be finite or infinite. Various topics covered under discrete mathematics include set theory, graph theory, combinations, logic, information theory, probability, geometry and algebraic structures. The notations and concepts of discrete mathematics are beneficial in analyzing and expressing objects and issues in the field of computer science, including programming languages, automated theorem proving, computer algorithms, software development and cryptography. Most of the topics introduced in this book cover new techniques and the applications of discrete mathematical structures. It will serve as a valuable source of reference for graduate and postgraduate students. Coherent flow of topics, student-friendly language and extensive use of examples make this book an invaluable source of knowledge.

The information shared in this book is based on empirical researches made by veterans in this field of study. The elaborative information provided in this book will help the readers further their scope of knowledge leading to advancements in this field.

Finally, I would like to thank my fellow researchers who gave constructive feedback and my family members who supported me at every step of my research.

Jonas Hughes

1

Sets, Propositions Relations and Functions

1.1 Principle of Inclusion and Exclusion

A binary operation on a set A is a function $\circ : A \times A \to A$. Binary operations are usually denoted by special symbols such as:

$+, -, , / , \times , \circ , \cap , \cup$, or, and.

We often write $a \circ b$ rather than $\circ (a, b)$.

Definition of Set

Let $\circ$ be a binary operation on a set A. An element $e \in A$ is an identity element for $\circ$ if for all $a \in A$, $a \circ e = a = e \circ a$.

Theorem

Let $\circ$ be a binary operation on A. Suppose that e and f are both identities for $\circ$. Then e = f. In other words, if an identity exists for a binary operation, it is unique.

Proof of Theorem

Since for all $a \in A$, $e \circ a = a$,

We get in particular that $e \circ f = f$. Also, for every $a \in A$, $a \circ f = a$, hence $e \circ f = e$.

Thus, $e = e \circ f = f$.

Definition:

Let $\circ$ be a binary operation on A and suppose that e is its identity. Let x be an element of A . An inverse of x is an element $y \in A$ such that $x \circ y = e = y \circ x$.

Three Important Points about Binary Operations:

- The result of the operation must be an element of S. This fails; for example, for + on the set S={−1, 0, 1} (as 1 + 1 = 2 ∉S).
- The operation must be defined for all elements of S. This fails; for example for A*B = A−1BA on Mn(R) (as the matrix A−1 may not exist).
- The result of the operation must be uniquely determined. This fails; for example, if we set:

  a*b = c where c2 = ab on C (as for a=b= 2, c may be 2 or −2).

Proposition

A proposition is a statement or declarative sentence which may be either true or false, but not both. Let p be the preposition, the statement "it is not the case that p" is another preposition called negation of p. It is read as not p.

The Inclusion-exclusion Principle

The inclusion-exclusion principle generalizes the rule of sum to non-disjoint sets.

Very often, we need to calculate the number of elements in the union of certain sets. Assuming that we know the sizes of these sets and their mutual intersections, the principle of inclusion and exclusion allows us to do exactly that.

Suppose that we have two sets A, B. The size of the union is certainly at most |A| + |B|. This way we are counting twice all the elements in A ∩ B, the intersection of the two sets. To correct this, we subtract |A ∩ B| to obtain the following formula:

$$|A \cup B| = |A| + |B| - |A \cap B|$$

In general, the formula gets more complicated because we have to consider the intersections of multiple sets. The following formula is called as the principle of inclusion and exclusion.

Lemma

For any collection of finite sets $A_1, A_2, \ldots, A_n$, we have:

$$\left|\bigcap_{i=1}^{n} A_i\right| = \sum_{\theta \neq i I \subseteq |n|} (-1)^{[I]+1} \left|\bigcap_{i \in I}^{n} A_i\right|$$

Writing out the formula more explicitly, we get:

$$|A1\cup \ldots An| = |A1| + \ldots + |An| - |A1\cap A2| - \ldots$$
$$|An-1\cap An| + |A1\cap A2\cap A3| + \ldots$$

In other words, we add up the sizes of the sets, subtract intersections of pairs, add intersection of triples, etc. The proof of this formula is very short and elegant, using the notion of the characteristic function.

Proof:

Assume that $A_1, \ldots, A_n \subseteq X$. For each set A_i, define the "characteristic function" $f_i(x)$ where $f_i(x) = 1$ if $x \in A_i$ and $f_i(x) = 0$ if $x /\in A_i$. We consider the following formula:

$$F(x)=\prod_{i=1}^{n}(1-f_i(x))$$

Observe that this is the characteristic function of the complement of $S_n i = 1\ A_i$: it is 1 if F(x) is not in any of the sets A_i.

Hence,

$$\sum_{x\in X} F(x) = \left|X/\bigcup_{i=1}^{n} A_i\right| \qquad ...(1)$$

Now we write F(x) differently, by expanding the product into 2n terms:

$$F(x)\prod_{i=1}^{n}(1-f_i(x))=\sum_{I\subseteq|n|}(-1)^{[I]}\prod_{i\in I}^{n} f_i(x)$$

Observe that $\prod_{i\in I} f_i(x)$ is the characteristic function of $\bigcap_{i\in I} A_i$. Therefore, we get:

$$\sum_{x\in X} F(x) = \sum_{I\subseteq|n|}(-1)^{[I]}\sum_{x\in X}\prod_{i\in I}^{n} f_i(x)=\sum_{I\subseteq|n|}(-1)^{[I]}\left|\bigcap_{i\in I} A_i\right| \qquad ...(2)$$

By comparing (1) and (2), we see that:

$$\left|X/\bigcup_{i=1}^{n} A_i\right|=|X|-\left|\bigcup_{i=1}^{n} A_i\right|=\sum_{I\subseteq|n|}(-1)^{[I]}\left|\bigcap_{i\in I} A_i\right|$$

The first term in the sum here is $\left|\bigcap_{i\in\theta} A_i\right|=|X|$ by convention.

Example:

1. Let us assume that in a university with 1000 students, 200 students are taking a

course in mathematics, 300 are taking a course in physics and 50 students are taking both. Let us calculate the number of students who are taking at least one of those courses.

Solution:

Given:

University with 1000 students

200 students are taking a course in mathematic

300 are taking a course in physics

50 students are taking both

If U = Total set of students in the university

M = Set of students taking Mathematics

P = Set of students taking Physics, then:

$|M \cup P| = |M| + |P| - |M \cap P| = 300 + 200 - 50 = 450$ students are taking Mathematics or Physics.

For three sets, the following formula applies:

$$|A \cup B \cup C| = |A| + |B| + |C| - |A \cap B| - |A \cap C| - |B \cap C| + |A \cap B \cap C|,$$

And for an arbitrary union of sets:

$$|A_1 \cup A_2 \cup \cdots \cup A_n| = s_1 - s_2 + s_3 - s_4 + \cdots \pm s_n,$$

Where s_k = sum of the cardinalities of all possible k-fold intersections of the given sets.

Generalizations of the Principle

Consider a set S with $|S| = N$, and conditions $c_1, c_2, \ldots c_t$ satisfied by some of the elements of S. Principle of Inclusion and Exclusion provides a way to determine $N(\overline{c}_1 \overline{c}_2 \ldots \overline{c}_t)$ which is the number of elements in S that satisfy none of the t conditions. If $m \in Z+$ and $1 \leq m \leq t$. We now want to determine E_m, which denotes the number of elements in S that satisfy exactly m of the t conditions (At present we can obtain E_o).

We can write formulas such as:

$$E_1 = N(c_1 \overline{c}_2 \overline{c}_3 \ldots \overline{c}_t) + N(\overline{c}_1 c_2 \overline{c}_3 \ldots \overline{c}_t) + \ldots + N(\overline{c}_1 \overline{c}_2 \overline{c}_3 \ldots \overline{c}_{t-1} c_t).$$

and,

$$E_2 = N(c_1 c_2 \bar{c}_3 .... \bar{c}_t) + N(c_1 \bar{c}_2 c_3 .... \bar{c}_t) + ... + N(\bar{c}_1 \bar{c}_2 \bar{c}_3 .... \bar{c}_{t-2} c_{t-1} c_t).$$

and although these results do not assist us as much as we should like, they will be a useful starting place as we examine the Venn diagrams for the cases where t = 3 and 4.

For figure, where t = 3, we place a numbered condition beside the circle representing those elements of S satisfying that particular condition and we also number each of the individual regions shown. Then E_1 equals the number of elements in regions 2, 3 and 4.

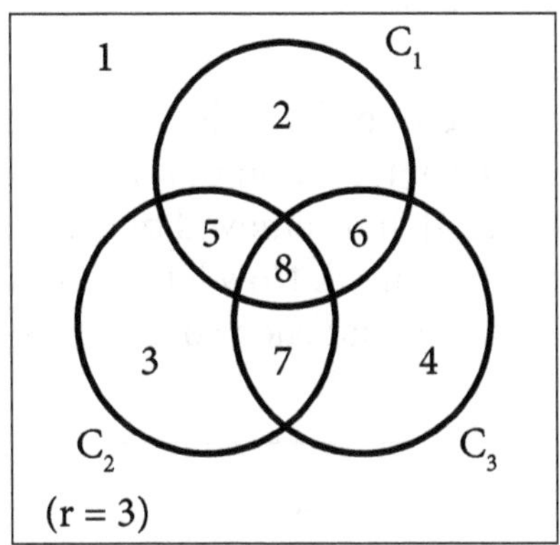

But we can also write:

$$E_1 = N(c_1) + N(c_2) + N(c_3) - 2[N(c_1c_2) + N(c_1c_3) + N(c_2c_3)] + 3N(c_1c_2c_3).$$

In $N(c_1)+ N(c_2) + N(c_3)$, we count the elements in regions 5, 6 and 7 twice and those in region 8 three times. In the next term, the elements in regions 5, 6 and 7 are deleted twice. We remove the elements in region 8 six times in $2\,[N(c_1c_2)+ N(c_1c_3)+ N(c_2c_3)]$, so we then add on the term $3N(c_1c_2c_3)$ and end up not counting the elements in region 8 at all. Hence, we have $E_1 = S_1 - 2S_2 + 3S_3 = S_1 - S_2 + S_3$.

When we turn to E_2, our earlier formula indicates that we want to count the elements of S in regions 5, 6 and 7.

From the Venn diagram:

$$E_2 = N(c_1c_2) + N(c_1c_3) + N(c_2c_3) - 3N(c_1c_2c_3) = S_2 - 3S_3 = S_2 - S_3.$$

and

$$E_3 = N(c_1c_2c_3) = S_3$$

In the figure, the conditions c_1, c_2, c_3 are associated with circular subsets of S, whereas c_4 is paired with the rather irregularly shaped area made up of regions 4, 8, 9, 11, 12, 13, 14 and 16.

For each $1 \leq i \leq 4$, E_i is determined as follows:

E_1 [regions 2, 3, 4, 5]:

$$E_1 = \left[N(c_1) + N(c_2) + N(c_3) + N(4)\right] - 2\left[N(c_1c_2) + N(c_1c_3) + N(c_1c_4) + N(c_2c_3) + N(c_2c_4) + N(c_3c_4)\right] + 3\left[N(c_1c_2c_3) + N(c_1c_2c_4) + N(c_1c_3c_4) + N(c_2c_3c_4) - 4N(c_1c_2c_3c_4)\right]$$

$$= S_1 - 2S_2 + 3S_3 - 4S_4$$

$$= S_1 - \binom{2}{1}S_2 + \binom{3}{2}S_3 - \binom{4}{3}S_4.$$

Taking an element in region 3, we find that it is counted once in E_1 and once in S_1 [in $N(c_3)$). Taking an element in region 3, we find that it is not counted in E_1. It is counted twice in S_1 [in both $N(c_2)$ and $N(c_3)$], but removed twice in $2S_2$ [for it is counted once in S_2 in $N(c_2c_3)$], so overall it is not counted. Now let us consider an element from region 12 and one from region 16 and show that each contributes a count of 0 to both sides of the formula for E_1.

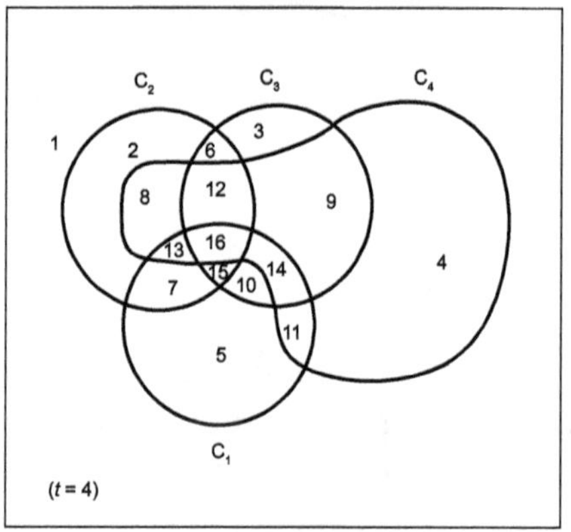

E_2 (Regions 6-11)

From the figure, $E_2 = S_2 - 3S_3 + 6S_4 = S_2 - S_3 + S_4$. For details on this formula, we examine the results in the table, where next to each summand of S2, S3 and S4 we list the regions whose elements are counted in determining that particular summand.

In calculating $S_2 - 3S_3 + 6S_4$ we find the elements in regions 6-11, which are precisely those that are to be counted for E_2.

S_2	S_3	S_4
$N(c_1c_2)$:7,13,15,16	$N(c_1c_2c_3)$:15,16	$N(c_1c_2c_3c_4)$:16
$N(c_1c_3)$:10,14,15,16	$N(c_1c_2c_4)$:13,16	
$N(c_1c_4)$:11,13,14,16	$N(c_1c_3c_4)$:14,16	

$N(c_2c_3)$:6,12,15,16	$N(c_2c_3c_4)$:12,16	
$N(c_2c_4)$:8,12,13,16		
$N(c_3c_4)$:9,12,14,16		

Finally, the elements for E_3 are found in regions 12-15 and $E_3 = S_3 - 4S_4 = S_3 - S_4$ the elements for E_4 are those in region 16, and $E_4 = S_4$. These results suggest the following theorem.

Theorem

For each $1 \le m \le t$, the number of elements in S that satisfy exactly m of the conditions $c_1, c_2, \ldots c_t$ is given by:

$$E_m = S_m - \binom{m+1}{1} S_{m+1} + \binom{m+2}{2} S_{m+2} - \ldots + (-1)^{t-m} \binom{t}{t-m} S_t$$

Proof

Let $x \in S$, consider the following three cases such as:

- x satisfies fewer than m conditions, it contributes 0 to both side.
- x satisfies exactly m of the conditions, it contributes 1 to both side (E_m and S_m).
- x satisfies r of the conditions, where $m \le r \le t$. Then r contributes nothing to E_m. Yet, it is counted $\binom{r}{m}$ times in S_m, $\binom{r}{m+1}$ times in S_{m+1} ,........ and $\binom{r}{r}$ times in S_r, but 0 times for any term beyond S_r.

So on the right side of the equation, x is counted $\binom{r}{m} - \binom{m+1}{1}\binom{r}{m+1} + \binom{m+2}{2}\binom{r}{m+2} - \ldots + (-1)^{r-m}\binom{r}{r-m}\binom{r}{r}$ times.

For $0 \le k \le r-m$:

$$\binom{m+k}{k}\binom{r}{m+k} = \frac{(m+k)!}{k!m!} \cdot \frac{r!}{(m+k)!(r-m-k)!}$$

$$= \frac{r!}{m!} \cdot \frac{1}{k!(r-m-k)!} = \frac{r!}{m!(r-m)!} \cdot \frac{(r-m)!}{k!(r-m-k)!}$$

$$= \binom{r}{m}\binom{r-m}{k}.$$

Consequently, on right-hand side of equation, x is counted:

$$\binom{r}{m}\binom{r-m}{0}-\binom{r}{m}\binom{r-m}{1}+\binom{r}{m}\binom{r-m}{2}-\ldots+(-1)^{r-m}\binom{r}{m}\binom{r-m}{r-m}$$

$$=\binom{r}{m}\left[\binom{r-m}{0}-\binom{r-m}{1}+\binom{r-m}{2}-\ldots+(-1)^{r-m}\binom{r-m}{r-m}\right]$$

$$=\binom{r}{m}[1-1]^{r-m}=\binom{r}{m}\cdot 0=0 \text{ times.}$$

And the formula is verified.

1.1.1 Mathematical Induction and Oppositions

Mathematical induction is one of the techniques which can be used to prove variety of mathematical statements which are formulated in terms of n, where n is a positive integer.

The Principle of Mathematical Induction

Let P(n) be a given statement involving the natural number n, such that:

- The statement is true for n = 1, i.e., P(1) is true (or true for any fixed natural number).
- If the statement is true for n = k (where k is a particular but arbitrary natural number), then the statement is also true for n = k+ 1, i.e., truth of P(k) implies the truth of P(k+ 1). Then, P(n) is true for all natural numbers n.

Problems

1. Using Mathematical induction, let us show that $\sum_{k=1}^{n} k^2 = \frac{n(n+1)(2n+1)}{6}$.

Solution:

Given:

$$\sum_{k=1}^{n} k^2 = \frac{n(n+1)(2n+1)}{6}.$$

Let $P(n): 1^2+2^2+\ldots+n^2=\frac{n(n+1)(2n+1)}{6}$ ...(1)

When n = 1,

$$P(1)=1^2=\frac{1(2)(3)}{6}.$$

So P(1) is true.

The basic step is valid.

Let us assume, $P(k): 1^2 + 2^2 + \ldots + k^2 = \frac{k(k+1)(2k+1)}{6}$ ...(2)

Now $P(k+1): 1^2 + 2^2 + \ldots + k^2 + (k+1)^2$

$$= \frac{k(k+1)(2k+1)}{6} + (k+1)^2 \text{ by } (2)$$

$$= \frac{k(k+1)(2k+1) + 6(k+1)^2}{6}$$

$$= \frac{(k+1)\left[2k^2 + 7k + 6\right]}{6}$$

$$= \frac{(k+1)(k+2)(2k+3)}{6}$$

$$= \frac{(k+1)\left[(k+1)+1\right]\left[2(k+1)+1\right]}{6}$$

i.e., p(k + 1) is True.

Hence, the inductive step is also true. Hence, by mathematical induction, P(k + 1) is true whenever p(k) is true.

$$\Rightarrow P(n) = \sum_{k=1}^{n} k^2 = \frac{n(n+1)(2n+1)}{6} \text{ is true.}$$

2. By mathematical induction, let us prove that for all $n \geq 1$, $n^3 + 2n$ is a multiple of 3.

Solution:

Given:

Let P(n): $n^3 + 2n$ is a multiple of 3.

P(1): $1 + 2 = 3$ is a multiple of 3.

Assume that P(k) is true i.e., $P(k) = k^3 + 2k$ is a multiple of 3 ...(1)

Now,

$$P(k+1) = (k+1)^3 + 2(k+1)$$

$$= k^3 + 3k^2 + 3k + 1 + 2k + 2$$

$$= \left(k^3 + 2k\right) + 3k^2 + 3k + 3$$

$$= \left(k^3 + 2k\right) + 3\left(k^2 + k + 1\right)$$

= Multiple of 3 + Multiple of 3 by 1)

= Multiple of 3.

i.e., P(k + 1) is true, whenever P(k) is true.

∴ By mathematical induction, $P(n) = n^3 + 2n$ is multiple of 3 for all $n \geq 1$.

3. Using the principle of mathematical induction, let us prove that $1^2 + 2^2 + 3^2 + \ldots\ldots + n^2 = (1/6)\{n(n + 1)(2n + 1\}$ for all $n \in N$.

Solution:

Given:

$$1^2 + 2^2 + 3^2 + \ldots\ldots + n^2 = (1/6)\{n(n + 1)(2n + 1\} \text{ for all } n \in N.$$

Let the given statement be P (n). Then:

$$P(n): 1^2 + 2^2 + 3^2 + \ldots\ldots + n^2 = (1/6)\{n(n + 1)(2n + 1)\}.$$

Putting n =1 in the given statement, we get:

$$\text{LHS} = 1^2 = 1 \text{ and RHS} = (1/6) \times 1 \times 2 \times (2 \times 1 + 1) = 1.$$

Therefore, LHS = RHS.

Thus, P(1) is true.

Let P(k) be true. Then:

$$P\left(k\right): 1^2 + 2^2 + 3^2 + \ldots\ldots + k^2 = (1/6)\left\{k\left(k + 1\right)\left(2k + 1\right)\right\}.$$

Now, $1^2 + 2^2 + 3^2 + \ldots\ldots\ldots + k^2 + \left(k + 1\right)^2$

$$= (1/6)\ \{k\left(k + 1\right)\left(2k + 1\right) + \left(k + 1\right)^2$$

$$= (1/6)\left\{\left(k + 1\right).\left(k\left(2k + 1\right) + 6\left(k + 1\right)\right)\right\}$$

$$= (1/6)\{\left(k + 1\right)\left(2k^2 + 7k + 6\}\right)$$

$$= (1/6)\{(k + 1)(k + 2)(2k + 3)\}$$

$$= 1/6\{(k + 1)(k + 1 + 1)[2(k + 1) + 1]\}$$

$$\Rightarrow P(k + 1): 1^2 + 2^2 + 3^2 + ... + k^2 + (k+1)^2$$

$$= (1/6)\{(k + 1)(k + 1 + 1)[2(k + 1) + 1]\}$$

⇒ P(k + 1) is true, whenever P(k) is true.

Thus, P(1) is true and P(k + 1) is true, whenever P(k) is true.

Hence, by the principle of mathematical induction, P(n) is true for all n ∈N.

4. Using the principle of mathematical induction, let us prove that 1·2 + 2·3 + 3·4 +.....+ n(n + 1) = (1/3){n(n + 1)(n + 2)}.

Solution:

Given:

$$1\cdot 2 + 2\cdot 3 + 3\cdot 4 + ... + n(n + 1) = (1/3)\{n(n + 1)(n + 2)\}.$$

Let the given statement be P (n). Then:

$$P(n): 1\cdot 2 + 2\cdot 3 + 3\cdot 4 + ... + n(n + 1) = (1/3)\{n(n + 1)(n + 2)\}.$$

Thus, the given statement is true for n = 1, i.e., P (1) is true.

Let P(k) be true. Then:

$$P(k): 1\cdot 2 + 2\cdot 3 + 3\cdot 4 + ... + k(k + 1) = (1/3)\{k(k + 1)(k + 2)\}.$$

Now, $1\cdot 2 + 2\cdot 3 + 3\cdot 4 + ... + k(k + 1) + (k + 1)(k + 2)$

$$= (1\cdot 2 + 2\cdot 3 + 3\cdot 4 + ... + k(k + 1)) + (k + 1)(k + 2)$$

$$= (1/3)\ k(k + 1)(k + 2) + (k + 1)(k + 2)\ [\text{using (i)}]$$

$$= (1/3)\ [k(k + 1)(k + 2) + 3(k + 1)(k + 2)$$

$$= (1/3)\{(k + 1)(k + 2)(k + 3)\}$$

$$\Rightarrow P(k+1): 1\cdot 2 + 2\cdot 3 + 3\cdot 4 + \ldots\ldots + (k+1)(k+2)$$

$$= (1/3)\{k+1)(k+2)(k+3)\}$$

⇒ P(k + 1) is true, whenever P(k) is true.

Thus, P(1) is true and P(k + 1)is true, whenever P(k) is true.

Hence, by the principle of mathematical induction, P(n) is true for all values of ∈ N.

5. Using the principle of mathematical induction, let us prove that $1 \cdot 3 + 3 \cdot 5 + 5 \cdot 7 + \ldots + (2n - 1)(2n + 1) = (1/3)\{n(4n^2 + 6n - 1)$.

Solution:

Given:

$$1\cdot 3 + 3\cdot 5 + 5\cdot 7 + \ldots\ldots + (2n-1)(2n+1) = (1/3)\{n(4n^2 + 6n - 1).$$

Let the given statement be P(n). Then:

$$P(n): 1\cdot 3 + 3\cdot 5 + 5\cdot 7 + \ldots\ldots + (2n-1)(2n+1) = (1/3)n(4n^2 + 6n - 1).$$

When $n = 1$, LHS $= 1\cdot 3 = 3$ and RHS $= (1/3) \times 1 \times (4 \times 12 + 6 \times 1 - 1) =$ × ×

$$= \{(1/3) \times 1 \times 9\} = 3.$$

$\therefore$ LHS = RHS.

Thus, P(1) is true.

Let P(k) be true. Then:

$$P(k): 1\cdot 3 + 3\cdot 5 + 5\cdot 7 + \ldots\ldots + (2k-1)(2k+1) = (1/3)\{k(4k^2 + 6k - 1) \quad \ldots(1)$$

Now,

$$1\cdot 3 + 3\cdot 5 + 5\cdot 7 + \ldots\ldots\ldots + (2k-1)(2k+1) + \{2k(k+1)-1\}\{2(k+1)+1\}$$

$$= \{1\cdot 3 + 3\cdot 5 + 5\cdot 7 + \ldots\ldots\ldots\ldots + (2k-1)(2k+1)\} + (2k+1)(2k+3)$$

$$= (1/3)\, k(4k^2 + 6k - 1) + (2k+1)(2k+3) \; [\text{using (i)}]$$

$$= (1/3)\left[(4k^3 + 6k^2 - k) + 3(4k^2 + 8k + 3)\right]$$

$$= (1/3)\left(4k^3 + 18k^2 + 23k + 9\right)$$

$$= (1/3)\left\{(k + 1)\left(4k^2 + 14k + 9\right)\right\}$$

$$= (1/3)[k + 1)\left\{4k(k + 1)^2 + 6(k + 1) - 1\right\}]$$

$$\Rightarrow P(k + 1): 1\cdot 3 + 3\cdot 5 + 5\cdot 7 + ... + (2k + 1)(2k + 3)$$

$$= (1/3)\left[(k + 1)\left\{4(k + 1)^2 + 6(k + 1) - 1\right)\right\}\right]$$

⇒ P(k + 1) is true, whenever P(k) is true.

Thus, P(1) is true and P(k + 1) is true, whenever P(k) is true.

Hence, by the principle of mathematical induction, P(n) is true for all n ∈ N.

6. Using the principle of mathematical induction, let us prove that $1/(1 \cdot 2) + 1/(2 \cdot 3) + 1/(3 \cdot 4) + ... + 1/\{n(n + 1)\} = n/(n + 1)$.

Solution:

Given:

$$1/(1\cdot 2) + 1/(2\cdot 3) + 1/(3\cdot 4) + ... + 1/\left\{n(n + 1)\right\} = n/(n + 1)$$

Let the given statement be P(n). Then:

$$P(n): 1/(1\cdot 2) + 1/(2\cdot 3) + 1/(3\cdot 4) + ... + 1/\left\{n(n + 1)\right\} = n/(n + 1).$$

Putting n = 1 in the given statement, we get:

$$\text{LHS} = 1/(1\cdot 2) = \text{and RHS} = 1/(1 + 1) = 1/2.$$

$$\text{LHS} = \text{RHS}.$$

Thus, P(1) is true.

Let P(k) be true. Then:

$$P(k): 1/(1\cdot 2) + 1/(2\cdot 3) + 1/(3\cdot 4) + ... + 1/\left\{k(k + 1)\right\} = k/(k + 1) \quad ...(1)$$

Now $1/(1\cdot 2) + 1/(2\cdot 3) + 1/(3\cdot 4) + ... + 1/\left\{k(k + 1)\right\} + 1/\left\{(k + 1)(k + 2)\right\}$

$[1/(1\cdot2) + 1/(2\cdot3) + 1/(3\cdot4) + ... + 1/\{k(k+1)\}] + 1/\{(k+1)(k+2)\}$

$= k/(k+1)+1/\{ (k+1)(k+2)\}. \{k(k+2)\cdot$

$+ 1\}/\{(k+1)^2/[(k+1)(k+2)]$ using ...(ii)

$= \{k(k+2) + 1\}/\{(k+1)(k+2\}$

$= \{(k+1)^2\}/\{(k+1)(k+2)\}$

$= (k+1)/(k+2) = (k+1)/(k+1+1)$

$\Rightarrow P(k+1): 1/(1\cdot2) + 1/(2\cdot3) + 1/(3\cdot4) + ...$

$+ 1/\{ k(k+1)\} + 1/\{(k+1)(k+2)\}$

$= (k+1)/(k+1+1)$

⇒ P(k + 1) is true, whenever P(k) is true.

Thus, P(1) is true and P(k + 1)is true, whenever P(k) is true. Hence, by the principle of mathematical induction, P (n) is true for all n ∈ N.

1.2 Logical Connectives

Logic

The rules of logic give precise meaning to the mathematical statements. These rules are used to distinguish between valid and invalid mathematical arguments. A major goal of logic is how to understand and how to construct the correct mathematical arguments.

These rules are used in the design of computer circuits, the construction of computer programs, the verification of the correctness of the program and in many other ways. Connectives are used for making compound propositions.

The important ones are the following (p and q represent given propositions):

Name	Represented	Meaning
Conjunction	$P \wedge Q$	"p and q"
Negation	$\neg P$	"not p"

Disjunction	PVQ	"p or q (or both)"
Implication	$p \rightarrow q$	"if p then q"
Exclusive Or	P⊕Q	"either p or q, but not both"
Bi-Conditional	$p \leftrightarrow q$	"p if and only if q"

Connectives

In propositional logic, generally, we use five connectives which are OR (V), AND (Λ), Negation/ NOT (¬), Implication / if-then (→), If and only if (⇔).

OR (V)

The OR operation of two propositions A and B (written as A V B) is true if at least any of the propositional variable A or B is true. The truth table is as follows:

A	B	AVB
True	True	True
True	False	True
False	True	True
False	False	False

AND (Λ)

The AND operation of two propositions A and B (written as A Λ B) is true if both the propositional variable A and B is true. Its truth table is as follows:

A	B	AΛB
True	True	True
True	False	False
False	True	False
False	False	False

Negation (¬)

The negation of a proposition A (written as ¬A) is false when A is true, and is true when A is false. Its truth table is as follows:

A	¬A
True	False
False	True

Implication / if-then (→)

An implication A→B is False if A is true and B is false. The rest cases are true. Its truth table is as follows:

A	B	A→B
True	True	True
True	False	True
False	True	False
False	False	True

If and Only if (⇔)

A⇔B is bi-conditional logical connective which is true when p and q are both false or both are true.

The truth table is as follows:

A	B	A⇔B
True	True	True
True	False	False
False	True	False
False	False	True

Tautologies

A Tautology is a formula which is always true for every value of its propositional variables.

Example:

Prove $[(A \rightarrow B) \wedge A] \rightarrow B$ is a Tautology.

The truth table is as follows:

A	B	A→B	(A→B)→A	[(A→B)→A]→B
True	True	True	True	True
True	False	False	False	True
False	True	True	False	True
False	False	True	False	True

As we can see, every value of $[(A \rightarrow B) \wedge A] \rightarrow B$ is "True", it is a tautology.

Contradictions

A Contradiction is a formula which is always false for every value of its propositional variables.

Example:

Prove (A ∨ B) ∧ [(¬A) ∧ (¬B)] is a Contradiction.

The truth table is as follows:

A	B	AVB	¬A	¬B	(¬A) ∧ (¬B)	(A V B) ∧ [(¬A) ∧ (¬B)]
True	True	True	False	False	False	False
True	False	True	False	True	False	False
False	True	True	True	False	False	False
False	False	False	True	True	True	False

As we can see, every value of (A ∨ B) ∧ [(¬A) ∧ (¬B)] is "False", it is a contradiction.

Contingency

A Contingency is a formula which has both some true and some false values for every value of its propositional variables.

Example:

Prove (A ∨ B) ∧ (¬A) a Contingency.

The truth table is as follows:

A	B	AVB	¬A	(A V B) ∧ (¬A)
True	True	True	False	False
True	False	True	False	False
False	True	True	True	True
False	False	False	True	True

As we can see, every value of (A ∨ B) ∧ (¬A) has both "True" and "False", it is a contingency.

1.2.1 Conditionals and Bi-Conditionals

A conditional statement is a statement which is performed by if true or false. For example, if p and q are two propositions, "if p then q" is known as conditional statement or implication. A statement is called bi-conditional when it expresses the idea that the presence of some property is a necessary and sufficient condition for the presence of some other property.

Conditional and Bi-Conditional Statements

Conditional statements and bi-conditional statements of different propositions may be obtained by conjunction, disjunction and negation of propositions.

Conditional Statements

Two statements connected by the phrase 'If ..... Then', it is called conditional statements. If p, q are two statements forming the implication, 'If p then q', then we denote it by '$p \rightarrow q$', which means 'p implies q'.

The term "conditional statement" is generally used, whereas in functional programming, the terms "conditional expression" or "conditional construct" are preferred, because all these terms have distinct meanings.

p	q	$p \rightarrow q$
T	T	T
T	F	F
F	T	T
F	F	T

Bi-Conditional Statements

The term Bi-conditional means asserting that the existence or occurrence of one thing or an event depends on, and is dependent on, the existence or occurrence of another event as " A if and only if B". We use bi-conditional to combine two conditions in the formation of compound condition.

A Bi-conditional statement is a logical statement combining two statements which says that something is necessary and sufficient for something else which contains the term "if and only if".

A statement is called bi-Conditional, if it expresses that the existence of some property is a necessary and sufficient condition for the existence of some other property. Such a statement is phrased by saying "A if and only if B". The symbol used for expressing Bi-conditional is triple bar or double sided arrow, we use this symbol () to express bi-conditional.

Examples and Truth Table of Bi-Conditional Statements

1. "If you breathe you live and if you live then you breathe which implies you live if and only if you breathe"

"We pass any course if and only if we study".

This statement implies the following things:

- If we pass this course, then we studied (and)
- If we study, then we pass the course.

The following is the rule for Bi-conditional, which is expressed by using the following truth table:

A	B	AB
T	T	T
T	F	F
F	T	F
F	F	T

According to the above truth table, the bi-Conditional statement is true, if both A and B are true or if A and B are both false. If A is true and B is false or vice versa, then the bi-Conditional statement is false. Therefore, the bi-Conditional statement AB is equivalent to the statement ((AB)&(BA)) which means that each side implies the other side means that if a bi-Conditional statement is true, then both the sides hold true.

Example Problem on Bi-Conditional Statements

1. State whether the following is true or false.

$1+1=4$ if and only if Earth is a sky.

Solution:

Given:

$1+1=4$ if and only if Earth is a sky.

In the above statement, there are two questions:

A. $1+1=4$?

B. if Earth is a sky?

Since both the above statements are false, the bi-conditional statement $A \leftrightarrow B$ is true.

From the above discussion, we can say each of the following are equivalent to the bi-conditional statement $A \leftrightarrow B$.

- A if and only if B.
- A is necessary and sufficient for B.

- A is equivalent to B.

We can reverse A and B in the above three phrases.

1.2.2 Logical Equivalences

If R is the relation on the set of positive integers such that (a, b) ∈ R if $a^2 + b$ is even, then let us prove that R is an equivalence relation.

Solution:

Given:

(a, b) ∈ R if $a^2 + b$ is even,

Let $a^2 + a - a(a + 1)$ = even. Since a and (a + 1) are consecutive positive integers.

∴ (a, a) ∈ R.

Hence, R is reflexive.

When $a^2 + b$ is even, a and b must be both even or both odd.

In either case, $b^2 + a$ is even.

∴ (a, b) ∈ R implies (b, a) e R.

Hence, R is symmetric.

When a, b, c are even, $a^2 + b$ and $b^2 + c$ are even.

Also $a^2 - c$ is even.

When a, b, c are odd, $a^2 + b$ and $b^2 + c$ are even.

Also $a^2 + c$ is even.

Then (a, b) ∈ R and (b, c) ∈ R ⇒ (a, c) ∈ R

Hence, R is transitive.

i.e., R is reflexive, symmetric and transitive.

Hence, R is an equivalence relation.

1.2.3 Predicate Calculus

The predicate or propositional function is a statement having the variables. For instance, "x + 2 = 7", "X is American", "x1 < y1", "p is a prime number" are predicates.

The truth value of the predicate depends on the value assigned to its variables. For instance, if we replace x with 1 in the predicate "x + 2 = 7", we may obtain "1 + 2 = 7", which is false, but if we replace it with 5, we get "5 + 2 = 7", which is true.

We represent a predicate by a letter followed by the variables enclosed between the parenthesis.

P(x), Q(x, y), etc.

An example for P(x) is the value of x for which P(x) is true.

A counter example is the value of x for which P(x) is false.

So, 5 is an example for "x + 2 = 7", while 1 is a counter example.

Each variable in the predicate is assumed to belong to a universe of discourse, for instance in the predicate, "n is an odd integer", where 'n' represents an integer, so the universe of discourse of n is the set of all integers.

In "X is American", we might declare that X is a human being. So in this case, the universe of discourse is the set of all human beings.

Free and Bound Variables in Predicate Logic

A formula containing a part of the form (x) p(x) or (∃x) p(x). Such a part is called as x-bound part of the formula. Any variable appearing in an x-bound part of the formula is called bound variable. Otherwise, it is termed as free occurrence.

Example of Free and Bound Variable in Predicate Logic

Let (i) $(\exists x)P(x, y)$

Here x is bound variable, y is free variable:

(ii) $(\exists x)P(x) \wedge Q(y)$

Here x is bound variable, y is free variable:

Problems

1. Without constructing truth table, let us obtain the PDNF of (P ∧ Q) ∨ (⌉P ∧ R) ∨ (Q ∧ R).

Solution:

Given:

Let $S \Leftrightarrow (P \wedge Q) \vee (\urcorner P \wedge R) \vee (Q \wedge R)$

Which is sum of elementary product (DNF) to obtain PDNF.

$$S \Leftrightarrow [(P \wedge Q) \wedge (R \vee \neg R)] \vee [(\neg P \wedge R) \wedge (Q \vee \neg Q)]$$

$$\vee [(Q \wedge R) \wedge (P \vee \neg P)]$$

$$\Leftrightarrow (P \wedge Q \wedge R) \vee (P \wedge Q \wedge \neg R) \vee (\neg P \wedge Q \wedge R) \vee (\neg P \wedge \neg Q \wedge R)$$

$$\vee (P \wedge Q \wedge R) \vee (\neg p \wedge Q \wedge R)$$

$$S \Leftrightarrow (P \wedge Q \wedge R) \vee (P \wedge Q \wedge \neg R) \vee (\neg P \wedge Q \wedge R) \vee (\neg P \wedge \neg Q \wedge R).$$

Which is PDNF

2. Without using Truth Table, let us show that $P \to (Q \to P) \leftrightarrow \neg p \to (p \to Q)$.

Solution:

Given:

$$\text{LHS}: P \to (Q \to P) \Leftrightarrow \neg P \vee (\neg Q \vee Q)$$

$$\Leftrightarrow \neg P \vee (P \vee \neg Q)$$

$$\Leftrightarrow (\neg P \vee P) \vee \neg Q$$

$$\Leftrightarrow T \vee \neg Q \leftrightarrow T \ldots (1)$$

$$\text{RHS}: \neg P \to (P \to Q) \leftrightarrow P \vee (\neg P \vee Q)$$

$$\Leftrightarrow (\neg P \vee P) \vee \neg Q$$

$$\Leftrightarrow T \vee \neg Q \leftrightarrow T \ldots (2)$$

From (1) & (2),

$$P \to (Q \to P) \Leftrightarrow \neg P \to (P \to Q).$$

3. Without using truth table, let us find the PCNF and PDNF of $P \to (Q \wedge P) \wedge (\neg P \to (\neg Q \wedge \neg R))$.

Solution:

Given:

$$\text{Let } S \Leftrightarrow p \to (Q \wedge P) \wedge (\neg p \to (\neg Q \wedge \neg R))$$

$$\Leftrightarrow (\neg P \vee (Q \wedge P)) \wedge (P \vee (\neg Q \wedge \neg R))$$

$$\Leftrightarrow [(\neg P \vee Q) \wedge (\neg P \vee P)] \wedge [(P \vee \neg Q) \wedge (P \vee \neg R)]$$

$$\Leftrightarrow (\neg P \vee Q) \wedge T \wedge (P \vee \neg Q) \wedge (P \vee \neg R)$$

$$\Leftrightarrow (\neg P \vee Q) \wedge (P \vee \neg Q) \wedge (P \vee \neg R)$$

Which is product of elementary sum (i.e., CNF).

To obtain PCNF:

$$\Leftrightarrow [(\neg P \vee Q) \vee (R \wedge \neg R)] \wedge [(P \vee \neg Q) \vee (R \wedge \neg R)] \wedge [(P \vee \neg R) \vee (Q \wedge \neg Q)]$$

$$\Leftrightarrow (\neg P \vee Q \vee R) \wedge (\neg P \vee \neg Q \vee \neg R) \wedge (P \vee \neg Q \vee R) \wedge (P \vee \neg Q \vee \neg R)$$

$$\wedge (P \vee Q \vee \neg R) \wedge (P \vee \neg Q \vee \neg R).$$

$$\text{PCNF of } S \Leftrightarrow (P \vee Q \vee \neg R) \wedge (P \vee \neg Q \vee R) \wedge (P \vee \neg Q \vee \neg R)$$

$$\wedge (\neg P \vee Q \vee R) \wedge (\neg P \vee Q \vee \neg R) \ldots (1)$$

To obtain PDNF, now:

PCNF of $\neg S \Leftrightarrow$ [Product of remaining maximum terms in (1)]

$$\text{PCNF of } \neg S \Leftrightarrow (P \vee Q \vee R) \wedge (\neg P \vee \neg Q \vee R) \wedge (\neg P \vee \neg Q \vee \neg R).$$

Now PDNF of S $\Leftrightarrow \neg$ [PCNF of $\neg$ S]

$$\Leftrightarrow \neg [(P \vee Q \vee R) \wedge (\neg P \vee \neg Q \vee R) \wedge (\neg P \vee \neg Q \vee \neg R)]$$

$$\Leftrightarrow (\neg P \wedge \neg Q \wedge \neg R) \vee (P \wedge Q \wedge \neg R) \vee (P \wedge Q \wedge R)$$

Which is PDNF of S.

5. Let us determine how many rows are needed for the truth table of the formula $(P \wedge \neg q) \leftrightarrow ((\neg r \wedge s) \rightarrow t)$.

Solution:

Given:

$$(P \wedge \neg q) \leftrightarrow ((\neg r \wedge s) \rightarrow t)$$

If there are n distinct components in a statement, the corresponding truth table will consist of 2^n rows corresponding to 2^n possible combinations. Here, (p, q, r, s, t) are 5 distinct components in this statement.

Number of rows are needed for the truth table = $2^5 = 32$.

Construct the truth table for the formula:

p	Q	P∧Q	˥(P∧Q)	˥P	˥Q	˥P∨˥Q	(1)⇔(2)
T	T	T	F	F	F	F	T
T	F	F	T	F	T	T	T
F	T	F	T	T	F	T	T
F	F	F	T	T	T	T	T

$\neg(P\wedge Q)\leftrightarrow(\neg P\vee\neg Q)$.

6. Without constructing truth table, let us verify whether Q ∨ (P ∧ ¬ Q) ∨ (¬ ∨ P ∧ ¬Q) is a contradiction or tautology.

Solution:

Given:

$$\text{Let } S\Leftrightarrow Q\vee(P\wedge\neg Q)\vee(\neg P\wedge\neg Q)$$

$$\Leftrightarrow\left[Q\vee(P\wedge\neg Q)\right]\vee(\neg P\wedge\neg Q)p;\text{ Associative law}$$

$$\Leftrightarrow\left[(Q\vee P)\wedge(Q\vee\neg Q)\right]\vee\neg(P\wedge Q);\text{Distributive law}$$

$$\Leftrightarrow\left[(P\vee Q)\wedge T\right]\vee\neg(P\vee Q);$$

$$\Leftrightarrow(P\vee Q)\vee\neg(P\vee Q);(\text{Tautology low})$$

$$\Leftrightarrow T$$

Hence, $Q\vee(P\wedge\neg Q)\vee(\neg P\vee\neg Q)$ is Tautology.

7. Let us show that (¬ P ∧ (¬ Q ∧ R)) ∨ (Q ∧ R) ∨ (P ∧ R) ⇔ R.

Solution:

Proof:

$$\text{Let}(\neg P\wedge(\neg Q\wedge R))\vee(Q\wedge R)\vee(P\wedge R)\Leftrightarrow R$$

$\Leftrightarrow (\neg P \wedge \neg Q \wedge R) \vee [(Q \wedge R) \vee (P \wedge R]$; Associative law

$\Leftrightarrow (\neg P \wedge \neg Q \wedge R) \vee [(P \vee Q) \wedge R]$; Distributive law

$\Leftrightarrow [(\neg P \wedge \neg Q) \vee (P \vee Q)] \wedge R$; Distributive law

$\Leftrightarrow [\neg (P \vee Q) \vee (P \vee Q)] \wedge R$; DeMorgan's law

$\Leftrightarrow T \wedge R; P \vee \neg P \Leftrightarrow T$

$\Leftrightarrow R$; identity law.

Hence, $(\neg P \wedge (\neg Q \wedge R)) \vee (Q \wedge R) \vee (P \wedge R) \Leftrightarrow R$.

1.2.4 Quantifiers

1. Let us prove that $R \rightarrow S$ can be derived from the premises $P \rightarrow (Q \rightarrow S)$, $\neg R \vee P$ and Q.

Solution:

Given:

$P \rightarrow (Q \rightarrow S)$

$\neg R \vee P$

Q

Using rule CP, let us consider R as the additional premise. We will derive S from the premises $P \rightarrow (Q \rightarrow S)$;] R V P and Q.

Sl. No.	Step	Rule	Reason
1.	R	Additional premise	-
2.	] R v P	P	-
3.	$R \rightarrow P$	T(2)	$P \rightarrow Q \Rightarrow$] P V Q
4.	P	T(l)(3)	P, P-»Q => Q
5.	$P \rightarrow (Q \rightarrow S)$	P	-
6.	$(Q \rightarrow s)$	T(4)(5)	$P, P \rightarrow Q \Rightarrow Q$
7.	Q	P	-
8.	S	T(6)(7)	$P, P \rightarrow Q$ -> Q
9.	$R \rightarrow S$	CP	-

2. Let us show that d can be derived from the premises (a → b) Λ (a → c), ¬ (b Λ c), (d v a).

Solution:

Given:

$$(a \to b)\Lambda(a \to c),\neg(b\Lambda c),(dva)$$

Sl. No.	Statement	Reason
1	(a → b) Λ (a → c)	P
2	a → b	T, 1 and Simplification
3	a → c	T, 1 and Simplification
4	⌉b → ⌉a	T, 2 and Contrapositive
5	⌉c → ⌉a	T, 3 and Contrapositive
6	(⌉b V⌉ c) → ⌉a	T, 4 and 5
7	⌉(b V c) → ⌉a	T and Demorgan's Law
8	⌉(b V c)	P
9	⌉a	T, 7, 8 and modus ponens
10	(d V a)	P
11	(a V d)	T, 10, commutative
12	⌉a → d	T(11)
13	d	T, 9, 12 and modus ponens.

Nested Quantifiers

3. If A = {a, b}, B = {b, c}; C= {1, c}, then let us determine:

(1) A (B C)

(2) (A B) (A C)

(3) A (B C)

(4) (A B) (A C).

Solution:

Given:

$$A=\{a,b\},\ B=\{b,c\};\ C=\{1,c\}$$

$$(B\ C)=\{1,\ b,\ c\}$$

(1) $A(B\ C)=\{(a, 1), (a, b), (a, c), (b, 1), (b, b), (b, c)\}$.

(2) $A\ B=\{(a, b), (a, c), (b, b), (b, c)\}$

$A\ C=\{(a, 1), (a, c), (b, 1), (b, c)\}$.

$\therefore\ (A\ B)\ (A\ C)=\{(a, b), (a, c), (b, b), (b, c), (a, 1), (b, 1)\}$

(3) $B\ C=\{c\}$

$A\ (B\ C)=\{(a, c), (b, c)\}$.

(4) $(A\ B)\ (A\ C)=\{(a, c), (b, c)\}$

4. Let us verify the validity of the following argument. Lions are dangerous animals. There are Lions. Therefore, there are dangerous animals.

Solution:

Given:

Lions are dangerous animals. There are Lions. Therefore, there are dangerous animals.

Let L(x): x is a Lion.

D(x): x is a dangerous animal.

Premises

Lions are dangerous animals: $(x)\ (L(x) \rightarrow D(x))$

There are Lions: $(\exists x)\ L(x)$.

Conclusion

There are dangerous animals: $(\exists x)\ D(x)$.

No.	Step	Rule	Reason
1.	$(\exists x)\ L(x)$	Rule P	-
2.	$L(a)$	ES(1)	-
3.	$(x)(L(x)\rightarrow D(x))$	Rule P	-
4.	$L(a)\rightarrow D(a)$	US (3)	-

5.	D(a)	Rule T(2), (4)	P(x)> $P(x) \rightarrow Q(x) \Rightarrow Q(x)$
6.	$(\exists x)D(x)$	EG (5)	-

Hence, the given statements are valid.

1.2.5 Theory of Inference

Inference

When looking at proving equivalences, we were showing that expressions in the form $p \leftrightarrow q$ were tautologies and writing $p \equiv q$. But we don't always want to prove $\leftrightarrow$. Often we only need one direction.

In general, mathematical proofs are "show that p is true" and can use anything we know is true to do it. Basically, we want to know that [everything we know is true] $\rightarrow$ p is a tautology.

Then we know p is true.

We can use the equivalences we have for this.

Since they are tautologies $p \leftrightarrow q$, we know that $p \rightarrow q$.

But we can also look for tautologies of the form $p \rightarrow q$.

Here is a tautology that would be very useful for proving things:

$$((p \rightarrow q) \wedge p) \rightarrow q.$$

Rules of Inference

If we have an implication tautology that we would like to use to prove a conclusion, we can write the rule like this:

$$p \rightarrow q$$

$$\therefore \frac{p}{q}$$

This corresponds to the tautology $((p \rightarrow q) \wedge p) \rightarrow q$.

The ∴ symbol is therefore.

The first two lines are premises.

The last is the conclusion.

This inference rule is called modus ponens (or the law of detachment).

Here are the rules of inference that we can use to build arguments:

1. Modus Ponens

$$\therefore \begin{array}{c} p \\ p \to q \\ \hline q \end{array}$$

2. Modus Tollens

$$\therefore \begin{array}{c} \neg q \\ p \to q \\ \hline \neg q \end{array}$$

3. Hypothetical Syllogism

$$\therefore \begin{array}{c} p \to q \\ q \to r \\ \hline p \to r \end{array}$$

4. Disjunctive Syllogism

$$\therefore \begin{array}{c} p \lor q \\ \neg p \\ \hline q \end{array}$$

5. Addition

$$\therefore \begin{array}{c} p \\ \hline p \lor q \end{array}$$

6. Simplification

$$\therefore \begin{array}{c} p \land q \\ \hline p \end{array}$$

7. Conjunction

$$\therefore \begin{array}{c} p \\ q \\ \hline p \land q \end{array}$$

8. Resolution

$$\begin{array}{l} p \vee q \\ \neg p \vee q \\ \hline \therefore\ q \vee r \end{array}$$

- Using these rules by themselves, we can do some very boring (but correct) proofs.
- e.g. "If I am sick, there will be no lecture today;" "Either there will be a lecture today or all the students will be happy;" "The students are not happy."

Translate into logic as: $s \rightarrow \neg l$, $l \vee h$, $\neg h$.

Then we can reach a conclusion as follows:

- $l \vee h$ [hypothesis].
- $\neg h$ [hypothesis].
- l [disjunctive syllogism using (1) and (2)].
- $s \rightarrow \neg l$ [hypothesis].
- $\neg s$ [modus tollens using (3) and (4)].

Problems

1. Let us show that $Q \vee (P \wedge \rceil Q) \vee (\rceil P \wedge \rceil Q)$ is a tautology.

Solution:

Given:

$$Q \vee (P \wedge \rceil Q) \vee (\rceil p \wedge \rceil Q)$$

$\Leftrightarrow Q \vee ((P \wedge \rceil Q) \vee (\rceil P \wedge \rceil Q))$ (Associative Property)

$\Leftrightarrow Q \vee ((P \vee \rceil P) \wedge \rceil Q)$($\because$ Distributive Property)

$\Leftrightarrow Q \vee (T \wedge \rceil Q)$ $\because P \vee \rceil P \leftrightarrow T$

$\Leftrightarrow Q \vee \rceil Q$ $\because$ Identity law

$\Leftrightarrow T$ (Tautology)

Hence, $Q \vee (P \wedge \rceil Q) \vee (\rceil P \wedge \rceil Q)$ is a Tautology.

2. Let us show that $P \rightarrow (Q \rightarrow R)) \rightarrow ((P \rightarrow Q) \rightarrow (P \rightarrow R))$ is a tautology.

Solution:

Given:

$$P \rightarrow (Q \rightarrow R)) \rightarrow ((P \rightarrow Q) \rightarrow (P \rightarrow R))$$

Truth table for $P \rightarrow (Q \rightarrow R)) \rightarrow ((P \rightarrow Q) \rightarrow (P \rightarrow R))$

P	Q	R	P→Q	P→R	Q→R	P→(Q→R)	(P→Q) → (P → R)	(A)→(B)
						(A)	(B)	
T	T	T	T	T	T	T	T	T
T	T	F	T	F	F	F	F	T
T	F	T	F	T	T	T	T	T
T	F	F	F	F	T	T	T	T
F	T	T	T	T	T	T	T	T
F	T	F	T	T	F	T	T	T
F	F	T	T	T	T	T	T	T
F	F	F	T	T	T	T	T	T

The given compound proposition is a tautology (since all truth values of final column are true).

4. Using rules of inference, let us show that S ∨ R is tautologically implied by (P ∨Q) ∧ (P → R) ∧ (Q → S).

Solution:

Given:

Premise $(P \vee Q) \wedge (P \rightarrow R) \wedge (Q \rightarrow S)$ as premise conclusion S ∨ R).

SI. No.	Step	Rule	Reason
1.	P V Q	P	-
2.	]P → Q	T(1)	P → Q ⇔]P V Q
3.	Q →S	P	-
4.	]p → s	T(2)(3)	P → Q, Q → R ⇒ P → R
5.	P → R	P	-
6.	]R -]P	T(5)	P → Q ⇔]Q →]P
7.	]R → S	T(4)(6)	P → Q, Q → R ⇒ P → R
8.	R V S	T(7)	P → Q ⇔]P V Q

Hence, S ∨ R is tautologically implied by:

$$(P \vee Q) \wedge (P \rightarrow R) \wedge (Q \rightarrow S).$$

5. Using CP rule, let us show that (x)(P(x) →Q(x)) ⇔ (x)P(x) →(x)Q(x).

Solution:

Given:

$$(x)(P(x)\rightarrow Q(x))\Leftrightarrow(x)\ P(x)\rightarrow(x)Q(x)$$

Assume (x) P (x) as an additional premise together with the given premise $(x)(P(x)\rightarrow Q(x))$ to derive (x) Q(x), then by Rule CP. $(x)\ P(x)\rightarrow(x)Q(x)$ can be derived from the premise alone $(x)\ P(x)\rightarrow Q(x)$.

Sl. No.	Step	Rule	Reason
1.	(x)P (x)	Additional premise	-
2.	(x)(P(x) → Q(x))	P	-
3.	P(a)	U.S. (1)	(x)P (x) ⇒ P(a)
4.	P(a) → Q(a)	U.S.(2)	(x)P (x) ⇒ P (a)
5.	Q(a)	T(3)(4)	P, P → Q ⇒ Q
6.	(x)(Q(x))	U.G.(5)	P(a) ⇒ (x)P(x)
7.	(x)P(x)→(x)Q(x))	C.P.	

6. Let us show that:

$$(P\rightarrow Q)\wedge(R\rightarrow S),\ (Q\wedge M)\wedge(S\rightarrow N),\urcorner(M\wedge N)\text{ and }(P\rightarrow R)\Rightarrow\urcorner P$$

Solution:

Given:

$$(P\rightarrow Q)\wedge(R\rightarrow S),\ (Q\wedge M)\wedge(S\rightarrow N),\urcorner(M\wedge N)\text{ and }(P\rightarrow R)\Rightarrow\urcorner P$$

By Indirect method, taking ⌉ (⌉ P) = P as additional premise.

S. No.	Step	Rule	Reason
1	P	Additional premise	-
2	P → R	P	-
3	(P → Q) ∧ (R → S)	P	-
4	(P → Q)	T(3)	P ∧ Q ⇒ P
5	(R → S)	T(3)	P ∧ Q ⇒ Q
6	R	T(1)(2)	P, P → Q ⇒ Q
7	S	T(6)(5)	P, P → Q ⇒ Q
8	(Q ∧ M) ∧ (S → N)	P	-

9	$(Q \wedge M)$	T(8)	$P \wedge Q \Rightarrow P$
10	$S \rightarrow N$	T(8)	$P \wedge Q \Rightarrow Q$
11	N	T(7)(11)	$P, P \rightarrow q \Rightarrow Q$
12	M	T(9)	$P \wedge Q \Rightarrow Q$
13	$M \wedge N$	T(11)(12)	$P,Q \Rightarrow P \wedge Q$
14	$\urcorner(M \wedge N)$	P	-
15	F	T(13),(14)	$P, Q \Rightarrow P \wedge Q$

Hence, $(P \rightarrow Q) \wedge (R \rightarrow S), (Q \wedge M) \wedge (S \rightarrow N), \urcorner(M \wedge N)$ and $(P \rightarrow R) \Rightarrow \urcorner P$.

1.2.6 Methods of Proof

1. Using indirect proof, let us show that D can be derived from $(A \rightarrow B) \wedge (A \rightarrow C)$, $\urcorner(B \wedge C)$ and $D \vee A$.

Solution:

Given:

$$(A \rightarrow B) \wedge (A \rightarrow C),\ \urcorner(B \wedge C) \text{ and } D \vee A$$

By indirect method, taking negation of conclusion ($\urcorner D$) as additional premise.

Sl.No.	Step	Rule	Reason
1.	$\urcorner d$	Additional Premise	-
2.	DVA	P	-
3.	A	T(1)(2)	Disjunctive Syllogism $\urcorner P, P \vee Q \Rightarrow Q$
4.	$(A \rightarrow B) \vee (A \rightarrow C)$	P	-
5.	$(A \rightarrow B)$	T(4)	$P \wedge Q \Rightarrow P \Rightarrow Q$
6.	$A \rightarrow C$	T(5)	$P \wedge Q \Rightarrow P \Rightarrow Q$
7.	B	T(3)(5)	$P, P \rightarrow Q \Rightarrow Q$
8.	C	T(3)(6)	$P, P \rightarrow Q \Rightarrow Q$
9.	$(B \wedge C)$	T(7)(8)	$P, Q \Rightarrow P \wedge Q$
10.	$\urcorner(B \wedge C)$	P	-
11.	F	T(9)(10)	$P \wedge \urcorner P \Leftrightarrow F$

Hence, by indirect method,

D can be derived from $(A \rightarrow B) \wedge (A \rightarrow C)$, $\urcorner(B \wedge C)$ and $D \vee A$.

1.3 Relations and Functions: Properties of Binary Relations

Reflexive

A binary relation R in a set X is reflexive if, for every x Є X, x R x, that is (x, x) Є R or R is reflexive in X ó (x) (x Є X ⨀ x R x).

For example:

- The relation is reflexive in the set of real numbers.
- The set inclusion is reflexive in the family of all subsets of a universal set.
- The relation equality of set is also reflexive.
- The relation is parallel in the set lines in a plane.
- The relation of similarity in the set of triangles in a plane is reflexive.

Symmetric

A relation R in a set X is symmetric if for every x and y in X, whenever x R y, then y R x. i.e., R is symmetric in X ó (x)(y)(x ϵ X∧y ϵ X∧x R y ⨀ y R x}

For example:

- The relation equality of set is symmetric.
- The relation of similarity in the set of triangles in a plane is symmetric.
- The relation of being a sister is not symmetric in the set of all people. However, in the set females, it is symmetric.

Transitive

A relation R in a set X is transitive if, for every x, y and z are in X, R is whenever x R y and y R z, then x R z. (i.e.,) transitive in X ó (x) (y) (z) (x Є X^ y Є X^ z Є X^x Ry^y Rz ⨀ x Rz).

For example:

- The relations <, £, >, ³ and = are transitive in the set of real numbers.
- The relations Í, Ì, Ê, É and equality are also transitive in the family of sets.
- The relation of similarity in the set of triangles in a plane is transitive.

Irreflexive

A relation R in a set X is irreflexive if, for every x Є X , (x, x) iX.

For example:

- The relation < is irreflexive in the set of all real numbers.
- The relation proper inclusion is irreflexive in the set of all nonempty subsets of a universal set.
- Let X = {1, 2, 3} and S = {(1, 1), (1, 2), (3, 2), (2, 3), (3, 3)} is neither irreflexive nor reflexive.

Anti-symmetric

A relation R in a set x is anti-symmetric if, for every x and y in X, whenever x R y and y R x, then x = y. symbolically, (x)(y)(x ∈ X^ ∈ X^R y ^R x ∈ x = y).

For example:

- The relations £, ³ and = are anti -symmetric.
- The relation Í is anti -symmetric in set of subsets.
- The relation "divides" is anti- symmetric in set of real numbers.
- Consider the relation "is a son of" on the male children in a family. Evidently the relation is not symmetric, transitive and reflexive.
- The relation "is a divisor of "is reflexive and transitive but not symmetric on the set of natural numbers.
- Consider the set H of all human beings. Let r be a relation " is married to " R is symmetric.
- Let I be the set of integers. R on I is defined as a R b if a - b is an even number. R is an reflexive, symmetric and transitive.

1.3.1 Closure of Relations

The closure of a relation R with respect to a property is the intersection of all relations R containing R.

Property: The reflexive closure of a relation R on a set S is the union R ∪ {(s, s)| s ∈ S}

Example: The reflexive closure of proper divisibility is divisibility.

The reflexive closure of a relation could be represented digraphically by drawing a self-loop at each vertex that did not already have one. In the matrix representation, one could write 1's down the main diagonal.

Property: The symmetric closure of a relation R is the relation $R \cup R^{-1}$.

Property: The transitive closure of a relation R on a set S is the relation,

$$R^{*} = \bigcup_{j=1}^{\infty} R^{j}$$

Example: The relation parent of the transitive closure is proper ancestor of. The reflexive, transitive closure is ancestor of.

1.4 Wars Hall's Algorithm

To find the transitive closure of a relation R, sometime the method of computing R2, R3, ... is inefficient for large number of set. Wars hall's algorithm is an efficient method for finding transitive closure of a relation R.

Let R be a relation on a set

$A = \{a_1, a_2, ...a_n\}$

If $x_1, x_2, ...x_n$, y is a path in R, then the vertices other than x, y are known as interior vertices and xRx_1, x_1Rx_2, $....x_nRy$.

$x_1, x_2, ...x_n$ are interior vertices of the path. For $1 \le i \le n$, define a Boolean matrix. W_k has 1 in position i, j if and only if there is a path from a_i to a_j in R whose interior vertices, if from the set $\{a_1, a_2, ...a_k\}$.

Since any vertex must come from the set $\{a_1, a_2, ... a_k\}$, it follows that the matrix W_n has 1 in position i, j if and only if some path in R connects a with $\{a_1, a_2, ...a_k\}$.

Hence,

$$W_n = M_R{}^{*}$$

If we define $W_0 = M_R$, then we will have a sequence $W_0, W_1 ... W_n$ whose first term is M_R and last term is $M_R{}^{*}$.

Wars hall's algorithm gives a procedure to compute each matrix Wk from the previous matrix W_{k-1}. Beginning with the matrix of relation R, we proceed one step at a time, until we reach the matrix of R*, in n steps. The matrices W_k, being different from powers of the matrix M_R, result in a considerable saving of steps in the computation of the transitive closure of relation R.

Suppose $W_{k-1} = [U_{jj}]$ and $W_k = [v_{ij}]$. If $v_{ij} = 1$, there is a path from ai to aj whose interior vertices come from the set lab $\{a_1,a_2,...a_k\}$. If ak is not an interior vertex of this path, then all the interior vertices must come actually from $\{a_1,a_2,...a_{k-1}\}$. Hence, $u_{ij} = 1$.

If a_k is an interior vertex of the path, then we must have the situation.

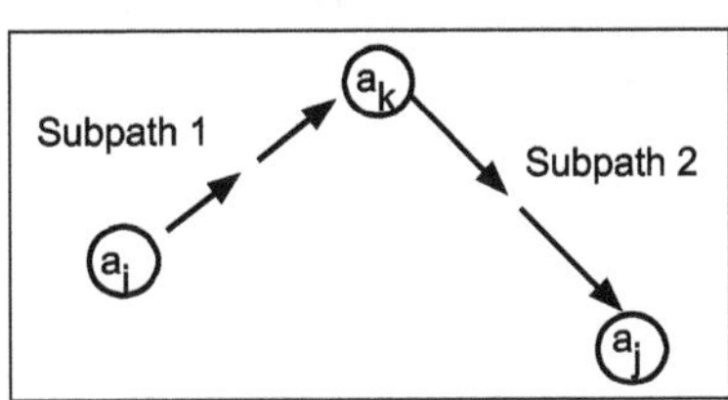

Since there is a sub path from ai to ak whose interior vertices come from $[a_1, a_2, ... a_{k-1}]$, we must have $u_{ik} = 1$, similarly $u_{kj} = 1$

Hence, $v_{ij} = 1$ if and only if:

- $u_{ij} = 1$
- $u_{ik} = 1$ and $u_{kj} = 1$

This is the basis of Wars hall's algorithm.

If W_{k-1} has 1 in position i, j then by (1), Wk will have 1 in position i, j. A new 1 can be added in position i, j of W_k if and only if column k of W_{k-1} has 1 in position i, and row k of W_{k-1} has 1 in position j.

Thus we have the following procedure for computing W_k from W_{k-1}.

Step 1: Transfer to W_k, all 1's in W_{k-1}.

Step 2: List the locations $p_1, p_2...$ in column k of W_{k-1} where the entry is 1, and locations q_1, q_2 ... in row k of W_{k-1}, where the entry is 1.

Step 3: Put 1's in all the positions p_i , q_j of W_k (if they are not already there).

Example:

Use Wars hall's algorithm to determine the transitive closure of R where $M_R = \begin{bmatrix} 1 & 0 & 1 \\ 0 & 1 & 0 \\ 1 & 1 & 0 \end{bmatrix}$ and A = {1, 2, 3}.

Solution:

Given:

$$W_0 = M_R = \begin{bmatrix} 1 & 0 & 1 \\ 0 & 1 & 0 \\ 1 & 1 & 0 \end{bmatrix}$$

A = {1, 2, 3}

n = 3

First, we find W1, so that k = 1, Wo has 1's in location 1 of column 1 and 3 of column 1. i.e., (1, 1) and (3, 1) and 1 of row 1 and 3 of row 1 i.e., (1, 1), (1, 3).

$$\begin{array}{ll} \text{i j} & \text{i j} \\ p_1:(1,1) & p_2=(3,1) \\ \text{i j} & \text{i j} \\ q_1=(1,1) & q_2=(1,3) \end{array}$$

Therefore, add (1, 1), (3, 3), (1, 3) in W1. Thus, W1 is just Wo with new in the position (1, 1), (3, 3), (1, 3).

$$W_1 = \begin{array}{c} \\ 1 \\ 2 \\ 3 \end{array}\begin{array}{c} \begin{array}{ccc} 1 & 2 & 3 \end{array} \\ \begin{bmatrix} 1 & 0 & 1 \\ 0 & 1 & 0 \\ 1 & 1 & 1 \end{bmatrix} \end{array}$$

Now to compute W_2, so that k_2, consider 2nd column and 2nd row.

p_1: (2, 2) p2: (3, 2)

q_1: (2, 2)

Therefore, add (2, 2), (3, 2) in W1 which are already 1 in W_1.

Hence,

$$W_2 = W_1 \begin{array}{c} \\ 1 \\ 2 \\ 3 \end{array}\begin{array}{c} \begin{array}{ccc} 1 & 2 & 3 \end{array} \\ \begin{bmatrix} 1 & 0 & 1 \\ 0 & 1 & 0 \\ 1 & 1 & 1 \end{bmatrix} \end{array}$$

p_1: (1, 3) p_2: (3, 3)

q_1: (3, 1) q_2: (3,2) q_3:(3,3)

Therefore, add (1, 1), (1, 2) (1, 3), (3, 1), (3, 2), (3, 3) in W_2 [if 1's are not already there]

$$W_3 = \begin{array}{c} \\ 1 \\ 2 \\ 3 \end{array}\begin{array}{c} \begin{array}{ccc} 1 & 2 & 3 \end{array} \\ \begin{bmatrix} 1 & 0 & 1 \\ 0 & 1 & 0 \\ 1 & 1 & 1 \end{bmatrix} \end{array}$$

Hence, R*= W3 is the transitive closure of R.

Therefore,

R* = {(1, 1), (1, 2), (1, 3), (2, 2), (3, 1), (3, 2), (3, 3)}

1.4.1 Equivalence Relations

A relation R in a set A is called an equivalence relation, if:

- R ϵ a for every a ϵ R i.e., R is reflexive.
- R ϵ b => b ϵ R ϵ a for every a, b ϵ A i.e., R is symmetric.
- R ϵ b and b ϵ R ϵ c => a ϵ R ϵ c for every a, b, c ϵ A, i.e., R is transitive.

For example:

- The relation equality of numbers on a set of real numbers.
- The relation being parallel on a set of lines in a plane.

Problems

1. Consider that the set T of triangles in a plane. And also a relation R in T as R= {(a, b)/ (a, b Є T and a is similar to b}. Let us also show that relation R is an equivalence relation.

Solution:

Given:

R = {(a, b)/ (a, b Є T and a is similar to b}

- A triangle a is similar to itself ⇒ a ϵ R ϵ a.
- If the triangle a is similar to the triangle b, then triangle b is similar to the triangle a i.e., a Є R Є b ⇒ b Є R Є a.
- If a is similar to b and b is similar to c, then a is similar to c (i.e.,) a Є R Є b and b Є R Є c => a Є R Є c. Hence, R is an equivalence relation.

2. Let us solve the, x = {1, 2, 3, ... 7} and R = {(x, y) / x - y is divisible by 3}. Let us show that R is an equivalence relation.

Solution:

Given:

For any a Є X, a- a is divisible by 3:

Hence, a Є R Є a, R is reflexive

For any a, b Є X, if a - b is divisible by 3, then b - a is also divisible by 3, R is symmetric.

For any a, b, c Є, if a Є R Є b and b Є R Є c, then a - b is divisible by 3 and b-c is divisible by 3. So that (a - b) + (b - c) is also divisible by 3, hence a - c is also divisible by 3. Thus R is transitive.

1.4.2 Partial Ordering Relations and Lattices

Partial and Total Ordering

A binary relation R in a set P is known as partial order relation or partial ordering in P if R is reflexive, anti-symmetric and transitive.

A partial order relation is denoted by the symbol £. If £ is a partial ordering on P, then the ordered pair (P, £) is called a partially ordered set or a potes.

Let R be the collection of real numbers. The relation "less than or equal to" or O, is a partial ordering on R. Let X be a set and r(X) be its power set. The relation subset, Í on X is partial ordering. Let S_n be the set of divisors of n. The relation D means "divides" on S_n, is partial ordering on S_n.

In a partially ordered set (P, £), an element y Î P is said to cover an element x Î P if x <y and if there does not exist any element z Î P such that x £ z and z £ y i.e., y covers x Û (x < y Ù (x £ z £ y Þ x =z Ú z = y)).

A partial order relation £ on a set P can be represented by means of a diagram called as a Hasse diagram or partial order set diagram of (P, £). In such a diagram, each element is represented by a small circle or a dot. The circle for x Î P is drawn below the circle for y Î P if x < y and a line is drawn between x and y if y covers x. If x < y but y does not cover x, then x and y are not connected directly by a single line. However, they are connected through one or more elements of P.

Total Orders

A partial order is a total or linear order iff for all x and y in the set, either x Є R Є y or y Є R Є x is true. In a totally ordered set, all elements are comparable.

Example:

The relation "less than or equal to" is a total order relation.

1.4.3 Chains and Anti-chains

Let (A, ≤) be a posset. A subset of A is known as chain if every pair of elements in the subset are related. The number of elements in a chain is termed as the length of the chain. A subset of A is known as anti-chain, if no two distinct elements in a subset are related.

Remark

Two elements a and b in a partially ordered set are said to be not comparable if a ≤ b (a is not related with b).

Totally Ordered Set

The word partial is used in defining a partial order in a set A because some elements in A need not be comparable. On the other hand, if every two elements in a partially ordered set A are comparable, then the partial order on A is called total order on A and the set A with relation ≤ is known as totally ordered set.

In other words, if A itself is a chain, the posset (A, ≤) is called a totally ordered set or linearly ordered set.

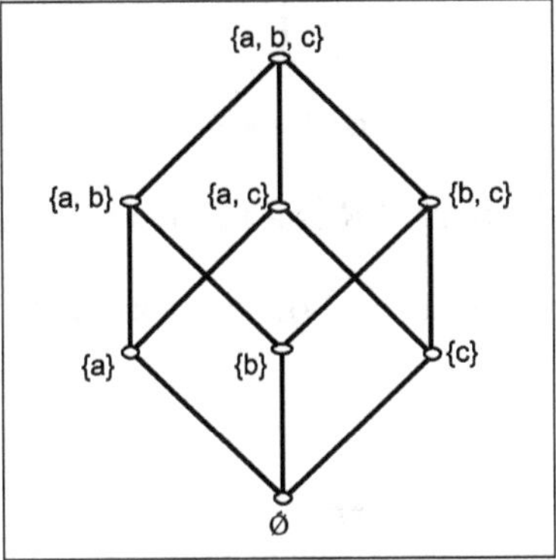

Example:

$S = \{a, b, c\}$ Pose $(P\ (S) \subseteq)$

Hasse Diagram

$\{ø\ \{a\}, \{a, b\}, \{a, b, c\}\}, \{ø, \{a\}, \{a, c\}, \{a, b, c\}\}$

$\{ø, \{b\}, \{a, b\}, \{a, b, c\}\}, \{ø, \{b\}, \{b, c\}, \{a, b, c\}\}$

$\{ø, \{c\}, \{a, c\}, \{a, b, c\}\}, \{ø, \{c\}, \{b, c\}, \{a, b, c\}\}$

$\{ø, \{a\}, \{a, b\}\}, \{a, \{a, c\}, \{a, b, c\}\}$ are the example of chains.

$\{\{a\}, \{b\}\}, \{\{a\}, \{c\}\}, \{\{b\}, \{c\}\}, \{\{a, b\}, \{a, c\}\}, \{\{b, c\}, \{a, c\}\}, \{\{a, b\}, \{b, c\}\}$

are the examples of antichains.

Totally Ordered Relation

Example:

1. A = {3, 9, 27, 81,...}

and aRb if a|b, then (A, R) is a totally ordered relation.

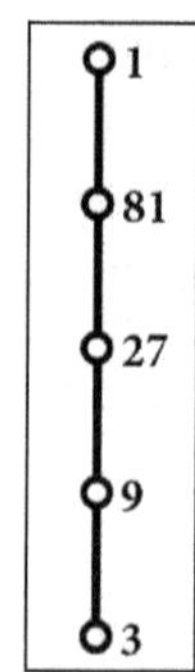

A itself is a chain.

2. N is the set of natural numbers and R is a relation defined as aRb if a ≤b.

Then, N is a chain and hence, N is a total ordered set with ≤

i.e., (N, ≤) in a totally ordered set.

Maximal and Minimal Elements

Let A be a non-empty set and ≤ is partial order relation on A. (A, ≤) is poset. An element a ∊ A is known as maximal element of A if there is no element c in A such that a ≤c. An element b ∊ A is known as minimal element of A if there is no element c in A such that c ≤b.

Example:

A = {2, 3, 5, 6, 10, 15, 30}

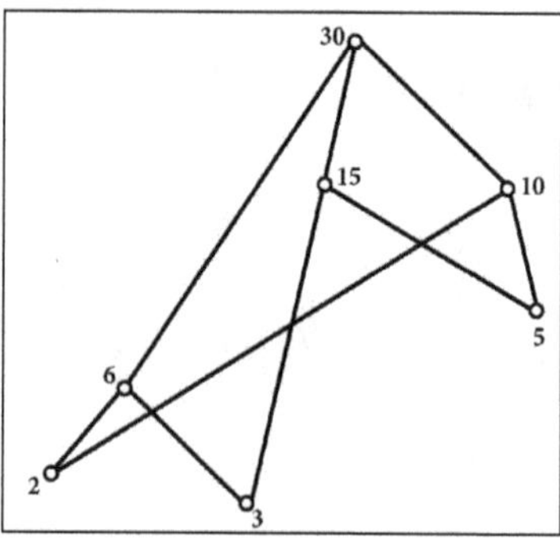

Hasse diagram of poset.

Maximal element is 30 and minimal elements are 2, 3, 5.

Upper Bounds and Lower Bounds

Let (A, ≤) be a poset. For elements a, b ∈ A, an element c ∈A is called upper bound of a and b if a≤ c and b ≤c.

c is known as least upper bound (lub) of a and b if c is an upper bound of a, b, and if there is no other upper bound d of a and b such that d ≤C. Lub is also known as supremum.

Similarly, an element e is said to be a lower bound of a and b if e≤ a and e ≤b; and e is known as greatest lower bound (glb) of a and b if there is no other lower bound f of a and b such that e ≤ f. glb is also known as infimum.

Example:

A = {2, 3, 5, 6, 10, 15, 30, 45}

Hasse Diagram

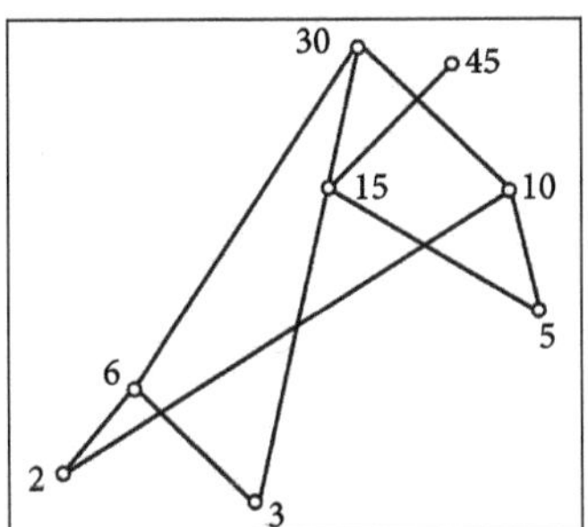

Here, 6 and 30 are the upper bounds of 2 and 3. 6 is least upper bound of 2 and 3. Similarly 15, 30 and 45 are the upper bounds of 3 and 5. 15 is the lub of 3 and 5. Also, 10 and 30 are upper bounds of 2 and 5. 10 in the lub of 2 and 5. Similarly 15, 3, 5 are the lower bounds of 30 and 45 in which 15 is the greatest lower bound.

Example:

{2, 3, 4, 6, 8, 12, 24, 36}.

Hasse Diagram

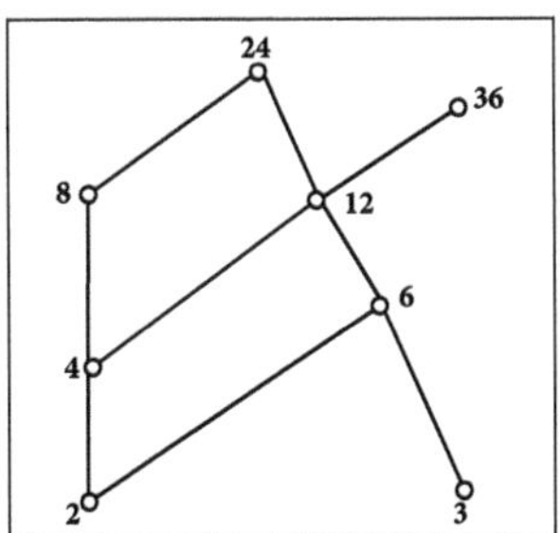

Upper bounds of 2 and 3 are 6, 12 and 24, 36 in which 6 is the least upper bound. Similarly lower bounds of 24 and 36 are 12, 4, 6, 2, 3 in which 12 is the greatest lower bound.

Example:

Let us draw the Hasse diagram of the following sets under the partial ordering relation 'divides' and indicate those which are chains:

1. {2, 4, 12, 24}
2. (1, 3, 5, 75, 30}

Solution:

Given:

1. A = (2, 4, 12, 24)

Hasse Diagram

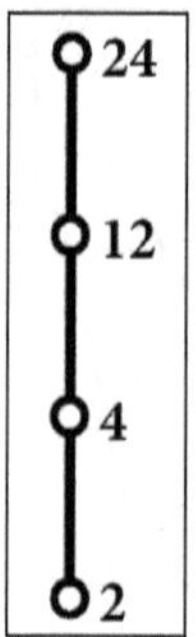

A itself is a chain. Hence (A, ≤) is totally ordered set or linearly ordered set.

2. S = {1, 3, 5, 15, 30)

Hasse Diagram

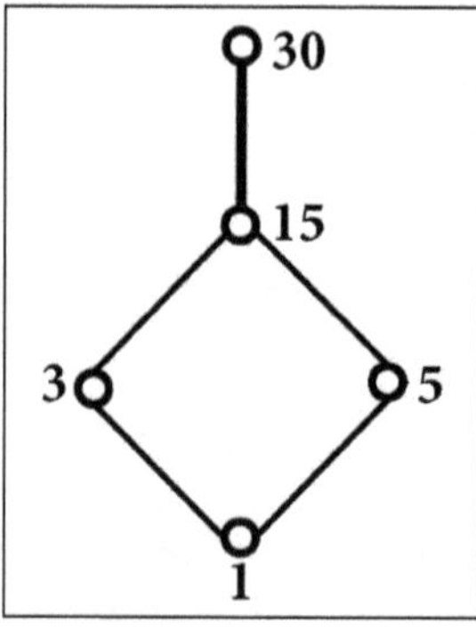

Chains are {1, 3, 15, 30} and {1, 5, 15, 30}

1.4.4 Functions

Let A be the set of 40 chairs in a classroom and B be the set of 30 students. The correspondence between the set A and B is "student sitting on chair." Then the correspondence f from A to B is a function or a mapping, if and only if:

- Every student is sitting on a chair.
- No student is sitting on two different chairs.

If these conditions are satisfied, f is called a function or mapping, and is denoted by f: A→ B.

Remarks

- If one student is standing, then f cannot be a function.
- If one student is occupying two different chairs still f cannot be a function.
  - i.e., f cannot be of the type one element is corresponding to many elements.
- Relation may be one-many but function cannot.
- Every function is a relation but every relation need not be a function.

Definition of Functions

Let A and B be two non-empty sets. Then a function or mapping f from the set A to the set B is a rule which assigns to each element a ϵ A to unique element b ϵ B.

We say that f maps element a to element b and that f maps set A to set B.

The notation denote that f maps a to b is,

$f(a) = b$ or $(a, b) \in f$.

[Remark: f is well defined if $f(a_1) = b$ and $f(a_1) = c$]

$b = c$

Representation by Diagram

Let the interiors of the two closed areas represent the sets A and B.

Let a_1, a_2, a_3... be the elements of A and b_1, b_2, b_3,..... be the elements of B.

The mapping or function f: A→ B is represented by means of arc of lines joining the points representing the elements of A to the elements of B.

- Every a ϵ A is joined to some b ϵ B.

- Two or more points in A may be joined to the same point in B (as a_1, a_2 are joined to b_2 in the figure).
- For mapping, two or more points of B cannot be joined to the same point in A.

Examples

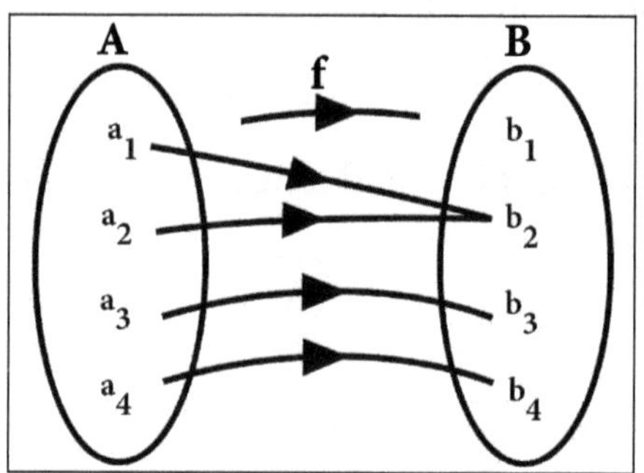

1. Let S_1 = (1, 2, 3)

S_2 = {p, q, r, S}

From the figure:

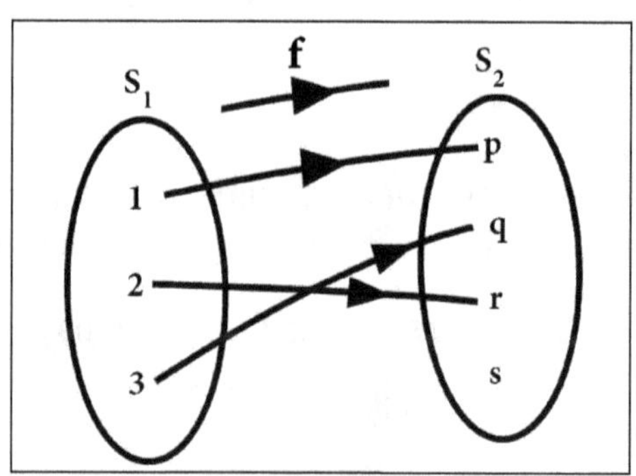

F: $S_1 \rightarrow S_2$ is a function it can also be denoted by $S_1 \xrightarrow{f.} S_2$

2. A = {x, y, z}, B= {1, 2, 3, 4}.

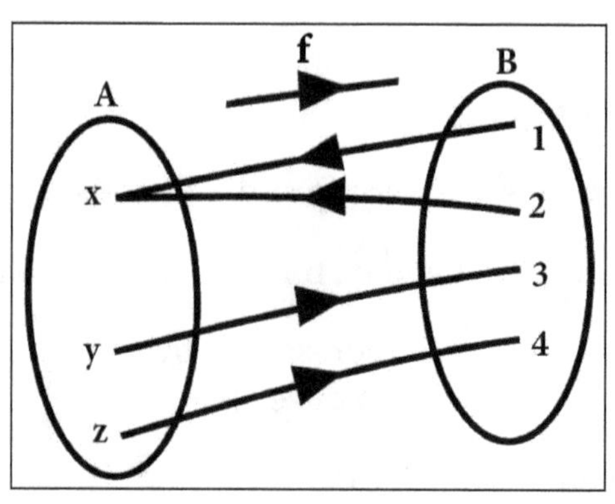

From the figure:

f is not a function from A to B as x in A is associated with two different elements of B.

3. A = {x, y, z}, B = {1, 2, 3, 4}

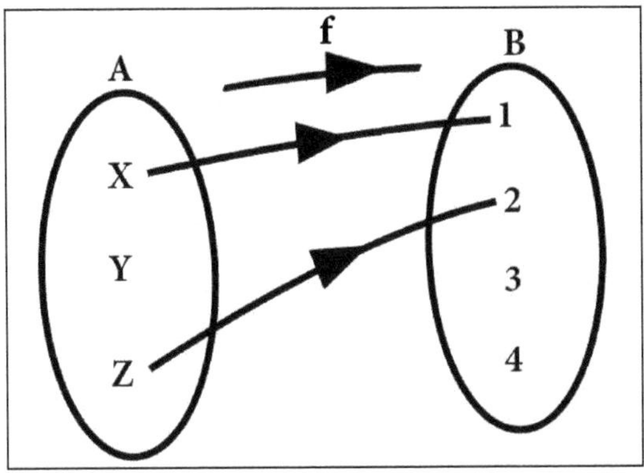

From the figure:

f is not a function as y from set A is not in correspondence with any element of set B.

4. A = {x, y, z, t}, = {1, 2, 3, 4, 5, 6}

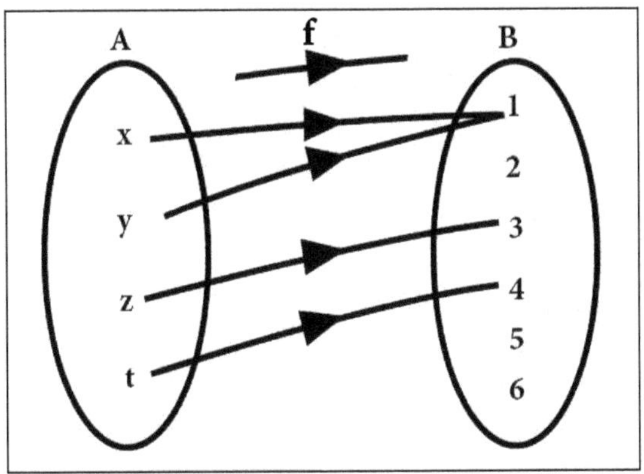

From the figure:

f is function from A to B as every element of set A is in correspondence with unique element of set B.

[Remark: f: A →B, f is a function from A to B, then two elements of A may correspond with one element of set B. Also, B may have elements with which no element of A is corresponding].

Image of Function f

If f is a function from A to B, i.e., f: A→ B

Then for f (a) = b, element b of B is called f image of element a of A and element a is called preimage of b.

Domain and Co-domain

If f is a function from A to B

i.e. f : A→ B. Then A set is known as domain set, B set is known as co-domain set.

Range of a Function

Range of (f) = (b : b ∊ B and f(a) = b for some a ∊ A).

In other words, range of f is the set of all images of the elements of A under f.

Example:

A ={x, y, z, t}, B = {1, 2, 3, 4, 5}

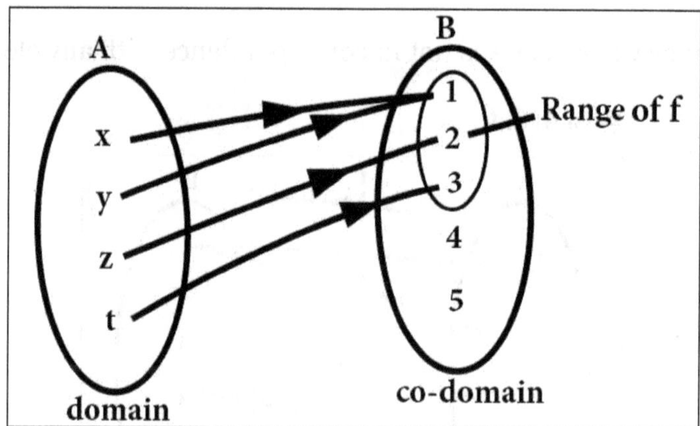

$f(x)=1$

$f(y)=1$

$f(z)=2$

$f(t)=3$

A is known as domain set B is known as co-domain set. Range of f is a subset of co-domain set. Range of f = {1, 2, 3} i.e., range of f is set of those elements of B which are associated with some elements of domain set A.

Elements of domain set are known as pre-image of f and elements of range are known as f-image of set A.

1.5 Composition of Functions

Injective/One-to-one Function

A function f: A→B is injective or one-to-one function if for every b ∈ B, there exists at most one a ∈ A such that f(s) =t.

This means a function f is injective if $a1 \neq a2$ implies $f(a1) \neq f(a2)$.

Example:

f: N→N, f(x)=5x is injective.

f: N→N, $f(x)=x^2$ is injective.

f: R→R, $f(x)=x^2$ is not injective as $(-x)^2=x^2$.

Surjective/On to Function

A function f: A→B is surjective (onto) if the image of f equals its range. Equivalently, for every b∈B, there exists some a ∈ A such that f(a)=b. This means that for any y in B, there exists some x in A such that y = f(x).

Example:

f: N→N, f(x) = x+2 is surjective.

f: R→ R, $f(x) = x^2$ is not surjective since we cannot find a real number whose square is negative.

Bijective/One-to-one Correspondent

A function f: A→B is bijective or one-to-one correspondent if and only if f is both injective and surjective.

Example:

Let us prove that a function f: R→R defined by f(x) = 2 x-3 is a bijective function.

Proof:

We have to prove this function is both injective and surjective.

If $f(x_1) = f(x_2)$, then $2x_1 - 3=2x_2 - 3$ and it implies that $x_1 = x_2$.

Hence, f is injective.

Here, 2x-3=y

So, x=(y+5)/3 which belongs to R and f(x) = y.

Hence, f is surjective.

Since f is both surjective and injective, we can say f is bijective.

Inverse of a Function

The inverse of a one-to-one corresponding function f: A→B, is the function g: B→A, holding the following property:

$$f(x)=y \Leftrightarrow g(y)=x$$

The function f is called invertible, if its inverse function g exists.

Example:

A function f: Z→Z, f(x) = x+5, is invertible since it has the inverse function g: Z → Z, g(x) = x−5.

A function f: Z→Z, f(x) = x^2 is not invertible since this is not one-to-one as $(-x)^2=x^2$.

Composition of Functions

Two functions f: A→B and g: B→C can be composed to give a composition gof. This is a function from A to C defined by (gof) (x) = g (f(x)).

Example:

$$f(x)=x+2 \text{ and } g(x)=2x$$

Let us determine (fog) (x) and (gof) (x).

Solution:

Given:

$$f(x)=x+2 \text{ and } g(x)=2x$$

$$(fog)(x)=f(g(x))=f(2x+1)=2x+1+2=2x+3$$

$$(gof)(x)=g(f(x))=g(x+2)=2(x+2)+1=2x+5$$

$$\text{Hence, } (fog)(x)\neq(gof)(x)$$

If f and g are one-to-one, then the function (gof) is also one-to-one.

If f and g are onto, then the function (gof)is also onto.

Composition always holds associative property, but does not hold commutative property.

1.5.1 Invertible Functions

A function f : A → B is called invertible if there exists a function g : B → A such that g ∘ f = 1A and f ∘ g = 1B.

Suppose f : A → B is a 1-1 correspondence. Then, by Proposition F11, f c is a function.

Consider the composite function f c∘ f. For a ∈ A we have (f c∘ f) = fc (f (a)) = a, by definition of converse. Thus, f c ∘ f = 1A.

Similarly, for b ∈ B we have (f ∘ f c)(b) = f c(f(b)) = b. Thus, f ∘ f c = 1B.

Therefore, a 1-1 and onto function is invertible.

Suppose f : A → B is invertible. Then the function g such that g ∘ f = 1_A and f ∘ g = 1_B is unique.

Proof:

Suppose g : B → A and h : B → A are functions such that g ∘ f = 1_A and f ∘ g = 1_B and h ∘ f = 1_A and f ∘ h = 1_B. We need to show g = h. Let b ∈ B. Then,

$g(b) = g(1_B(b)) = (g \circ 1_B)(b)=(g \circ (f \circ h))(b) = ((g \circ f) \circ h)(b) = (1_A \circ h)(b)=1_A(h(b)) = h(b)$. Thus, g = h. This completes the proof.

1.5.2 Recursive Functions

The term "recursive function" is commonly used informally to describe any function that is defined with recursion. There are many formal counterparts to this informal definition, several of which only differ in trivial respects.

Kleene (1952) defines a "partial recursive function" of nonnegative integers to be any function that is defined by a non-contradictory system of equations whose left and right sides are composed from:

- Function symbols (for example, f, g, h, etc.).
- Variables for nonnegative integers (for example, x, y, z , etc.).
- The constant 0.
- The successor function S(x) = x+1.

For example:

$$f(x,\ 0)=0 \qquad ...(1)$$

$$f(x,\ S(y))=g(f(x\ ,\ y),\ x) \qquad ...(2)$$

$$g(x,0)=x \qquad ...\ 3 \qquad ...(3)$$

$$g(x,\ S(y))=S(g\ (x,\ y)) \qquad ...(4)$$

Defines f(x, y) to be the function x y that computes the product of x and y.

Note that the equations may not uniquely find the value of f for every possible input and in that sense the definition is "partial". If the system of equations determines the value of f for every input, then the definition is said to be "total". When the term "recursive function" is used solitary, it is usually implicit that "total recursive function" is intended.

The set of functions that can be defined recursively in this manner is known to be equivalent to the set of functions computed by Turing machines and by the lambda calculus.

1.5.3 Pigeonhole Principle

Pigeonhole Principle states that if there are fewer pigeon holes than total number of pigeons and each pigeon is put in a pigeon hole, then there must be at least one pigeon hole with more than one pigeon. If n pigeons are put into m pigeonholes where n>m, there's a hole with more than one pigeon.

Problems

1. Let us show that f: R - {3} → R - {1} given by f(x) = (x - 2)/(x - 3) is a bisection and let us find its inverse.

Solution:

Given:

$f(x)=(x-2)/(x-3)$ is a bijection

To prove that f is bijection

i.e., TPT f is one-one and onto.

Let f (x) = f(y)

$$\Rightarrow \frac{x-2}{x-3}=\frac{y-2}{y-3}$$

$$\Rightarrow xy - 2y - 3x + 6 = xy - 2x - 3y + 6$$

$$\Rightarrow x = y$$

$\therefore$ f is one-one.

$$\text{Let } y = \frac{x-2}{x-3} \Rightarrow xy - 3y - x + 2 = 0.$$

$$\Rightarrow x\ (y\ -\ 1) = 3y - 2$$

i.e., $x = \dfrac{3y-2}{y-1}$ is pre-image of $y \in R\ - \{1\}$.

$\therefore$ f is onto

Hence, f is bijection.

Inverse function:

$$f^{-1}(y) = x = \frac{3y-2}{y-1}$$

$$f^{-1}(x) = \frac{3x-2}{x-1}.$$

2. Let us test whether the function f: Z → Z defined by f(x) = x^2 + 14x - 51 is an injective and a surjective.

Solution:

Given:

$f(x) = x^2 + 14x - 51$ is an injective.

If $f(x_2) = f(x_2)$ we have:

$$x_1^2 + 14x_1 - 51 = x_2^2 + 14x_2 - 51$$

$$\text{i.e.,} \left(x_1^2 - x_2^2\right) + 14\left(x_1 - x_2\right) = 0$$

$$\text{i.e.,}\ \left(x_1 - x_2\right)\ \left(x_1 + x_2 +\ 14\right) = 0$$

If we assume that $x_1 \neq x_2$. Then $f(x_1)$ can be equal to $f(x_2)$ provided that,

$$x_1 + x_2 +\ 14 = 0$$

For example, when $x_1 = -4, x_2 = -10$.

Thus, when $x_1 = -4$ and $x_2 = -10$ i.e., $x_1 \neq x_2$

$$f(x_1) = f(x_1) = -91$$

∴ f(x) is not injective (1 - 1).

If $y = f(x) = x^2 + 14x - 51$, we have $y = (x+7)^2 - 100$

i.e., $y + 100 = (x+y)^2$

∴When $y \in z$ and y less than - 100, x does not exist.

Hence, f(x) is not surjective (onto).

i.e., f(x) is neither injective nor surjective.

2

Numeric and Generating Functions, Recursive Relations and Algorithms

2.1 Discrete Numeric Functions

A function f from natural numbers set (with 0) to real numbers set R is known as discrete numeric function or numeric function. If $x : N \rightarrow \Re$ is a discrete numeric function, then $x_o, x_1, x_2, \ldots x_k, \ldots$ denote the values of the numeric function.

The numeric function x can be expressed as:

$$x \quad [x_0, x_1, x_2, \ldots\ldots\ldots, x_k, \ldots\ldots\ldots]$$

The numeric function is also known as sequence.

Examples:

1. $x_k = \begin{cases} 2^k & 0 \le k < 6 \\ k & k \ge 7 \end{cases}$

This discrete function can be rewritten as:

$$x = \{1, 2, 4, 8, 16, 32, 64, 7, 8, 9, \ldots.\}$$

2. $x_k = \begin{cases} k+4 & 0 \le k \le 5 \\ 3k-2 & k \ge 6 \end{cases}$

This discrete function can be rewritten as:

$$x = \{4, 5, 67, 8, 9, 16, 19, 22, 25, \ldots.\}$$

3. $x_k = \begin{cases} \frac{1}{3^k} & k \ge 4 \\ 0 & k < 4 \end{cases}$

$$x = \left\{0, 0, 0, 0, \frac{1}{3^4}, \frac{1}{3^5}, \frac{1}{3^6} \ldots.\right\}$$

Basic Operations on Numeric Functions

Addition

Let x and y be two numeric functions, then z = x + y is the addition of two numeric functions, where:

$$z_k = x_k + y_k$$

Multiplication

Multiplication of x and y is a numeric function say d, then d = xy, where:

$$d_k = x_k y_k$$

Scaling

If a is real number, then ax is numeric function and a is known as scaling factor.

Linearity

If a and b are real numbers, then t = ax + by is a numeric function, where:

$$tk = ax_k + by_k.$$

Examples:

1. $x_k = \frac{1}{3^k} \qquad k \geq 0$

Then,

$$x = \left\{1, \frac{1}{3}, \frac{1}{3^2}, \ldots\right\}$$

And,

$$5x = \left\{5, \frac{5}{3}, \frac{5}{3^2}, \ldots\right\}$$

$$\text{i.e } 5\frac{1}{3^k}$$

Where 5 is scaling factor.

2. Let $x_k = \frac{1}{4^k}, k \geq 0$

$$y_k = \begin{cases} 5 & 0 \le k \le 6 \\ k & k \ge 6 \end{cases}$$

Then,

$$z_k = x_k + y_k$$

$$= \begin{cases} \frac{1}{4^k} + 5 & 0 \le k < 6 \\ \frac{1}{4^k} + k & k \ge 6 \end{cases}$$

3. $x_k = \begin{cases} 4^k & 0 \le k < 3 \\ 2^k & k \ge 3 \end{cases}$

$$y_k = \begin{cases} 3^k & 0 \le k < 2 \\ k+5 & k \ge 2 \end{cases}$$

Then the product of x and y is z.

Where,

$$z_k = \begin{cases} 4^k \cdot 3^k & 0 \le k < 2 \\ 7 \cdot 4^2 & k = 2 \\ 2k \cdot (k+5) & k \ge 3 \end{cases}$$

4. $x_k = \begin{cases} 1 & 0 \le k < 5 \\ k & k \ge 6 \end{cases}$

$$y_k = \frac{1}{2^k} k \ge 0$$

Then,

$$Z = ax + by$$

Where,

$$z_k = \begin{cases} a \cdot 1 + b \cdot \frac{1}{2^k}, & 0 \le k < 5 \\ a \cdot k + b \cdot \frac{1}{2^k}, & k \ge 6 \end{cases}$$

Convolution of Two Numeric Functions x and y

Convolution is denoted by '*'

Hence, z=x*y is convolution of numeric function x and y and $z_k = X_0 y_k + x_1 y_{k-1} + x_2 y_{k-2} + ...x_k y_0$

$$= \sum_{n=0}^{k} x_n \, y_{k-n}$$

Example:

Let $X_k = 3^k \; k \geq 0$

$Y_k = 5^k \; k \geq 0$

then $z = x * y$

$$= \sum_{n=0}^{\infty} 3^n \, (5)^{k-n}$$

Finite difference of a numeric function x is a numeric function.

2.1.1 Generating Functions

The concept of generating functions is a powerful tool for solving counting problems. In counting problems, we are often interested in counting the number of objects of 'size n', which is denote by n a.

By varying n, we get different values of an. In this way, we get a sequence of real numbers a, a_1, a_2, ... from which we can define a power series.

$$G(x) = a_0 + a_1 x + a_2 x + ...$$

The above G(x) is the generating function for the sequence a_0, a_1, a_2, ...

Problems

1. Let us find the generating functions for the following sequences. In each case, try to simplify the answer.

(a)1, 1, 1, 1, 1, 1, 0, 0, 0, 0, ...

(b)1, 1, 1, 1, 1, ...

(c) $1, 3, 3, 1, 0, 0, 0, 0, \ldots$

(d) $C_0^{2005}, C_1^{2005}, C_2^{2005}, \ldots, C_{2005}^{2005}, 0, 0, 0, 0, \ldots$

Solution:

(a) The generating function is:

$$G(x)=1+1x+1x^2+1x^3+1x^4+1x^5+0x^6+0x^7+\ldots=1+x+x^2+x^3+x^4+x^5$$

We can apply the formula for the sum of a geometric series to rewrite $G(x)$ as $G(x)=\dfrac{1-x^6}{1-x}$.

(b) The generating function is:

$$G(x)=1+x+x^2+x^3+x^4+x^5$$

When $|x| = 1$, we can apply the formula for the sum to infinity of a geometric series to rewrite G(x) as $G(X)=\dfrac{1}{1-x}$.

(c) The generating function is $G(x)=1+3x+3x^2+1$, and of course, the binomial theorem enables us to simplify the answer as $G(x)=(1+x)^3$.

(d) The generating function is:

$G(X)=C_0^{2005}+C_0^{2005}X+C_0^{2005}X^2+\ldots+C_{2004}^{2005}X^{2004}+C_{2005}^{2005}X^{2005}$ and the binomial theorem once again enables us to simplify the answer as $G(x)=(1+x)^{2005}$.

When dealing with the computations of generating functions, we are particularly interested with whether the generating function can be written in closed form and whether we can find the coefficient of a certain power of x easily.

To write the generating function in 'closed form' means, in general, writing it in a 'direct' form without summation sign nor '?'. For instance $G(x) = 1 + x + x^2 + x^3 + x^4 + \ldots$ is not in the closed form.

The reason for trying to put a generating function in closed form is as follows:

- If we can find a generating function in closed form, the computations can be greatly simplified and easily carried out.
- In the more advanced theory of generating functions, we will find that certain combinations of problems correspond to certain operations on the generating functions.

On the other hand, we are interested in knowing the coefficient of a certain power of x because, it often refers to the number of objects of size n, which is usually the thing we wish to find in the counting problems.

Clearly, if a generating function is given in 'explicit form' such as:

$$G(x)=x+2x^2+3x^3+4x^4+.... \text{or}\, G(x)=\sum_{n=0}^{\infty}\frac{n-1}{2n+1}x^n$$

Then, finding a specific coefficient will be easy. However, if a generating function is given in the closed form ingenious tricks are sometimes required in determining certain coefficients.

2. Let us determine the coefficient of x^{2005} in the generating function

$$G(x)=\frac{1}{(1-x)^2(1+x)^2}.$$

Solution:

Given:

$$G(x)=\frac{1}{(1-x)^2(1+x)^2}.$$

$$\text{Let}\frac{1}{(1-x)^2(1+x)^2}=\frac{A}{1-x}+\frac{B}{(1-x)^2}+\frac{C}{1+x}+\frac{D}{(1+x)^2}$$

Upon simplification, the right hand side becomes:

$$\frac{(C-A)x^3(B+D-A-C)x^2+(A+2B-C-2D)x+(A+B+C+D)}{(1-x)^2(1+x)^2}$$

Comparing coefficients, we have:

$$\begin{cases} C-A & =0 \\ B+D-A-C & =0 \\ A+2B-C-2D & =0 \\ A+B+C+D & =1 \end{cases}$$

Solving, we get A=B=C=D=1/4, It follows that:

$$G(x)=\frac{1}{4}\left[(1-x)^{-1}+(1-x)^{-2}+(1+x)^{-1}+(1+x)^{-2}\right]$$

Thus, the coefficient of x^{2005} is $\frac{1}{4}\left(-C_{2005}^{-1}-C_{2005}^{-2}+C_{2005}^{-1}+C_{2005}^{-2}\right)=0$

3. Let us determine the generating functions of the following sequences in closed form.

(a) 1, 2, 3, 4, 5, 6, 7

(b) $0,1,-\frac{1}{2},\frac{1}{3},-\frac{1}{4},\ldots$

Solution:

Given:

(a) 1, 2, 3, 4, 5, 6, 7

(b) $0,1,-\frac{1}{2},\frac{1}{3},-\frac{1}{4},\ldots$

Formally, we can differentiate and integrate a power series term by term.

In other words, if:

$$G(x)=a_0+a_1x+a_2x^2+a_3x^3+\ldots.$$

Then,

$$G(x)=a_1+a_1x+2a_2x^2+3a_3x^3+\ldots.$$

The same is true for integration in place of differentiation.

(a) The generating function is:

$$G(x) = 1+2x+3x^2+4x^3+\ldots\ldots$$

$$=\frac{d}{dx}\left(x+x^2+x^3+x^4+\ldots\right)$$

$$=\frac{d}{dx}\left(\frac{x}{1-x}\right)$$

$$=\frac{1}{(1-x)^2}$$

We can verify the answer by expanding $\frac{1}{(1-x)^2}$ using the extended binomial theorem.

(b) The generating function is:

$$G(x)=x-\frac{1}{2}x^2+\frac{1}{3}x^3-\frac{1}{4}x^4+...$$

$$=\int(1-x+x^2-x^3+....)dx$$

$$=\int\frac{dx}{1+x}$$

$$=\ln(1+x)+C$$

To find the constant C, we put in x = 0 to get C = G (0). If we write:

$$G(x)=a_0+a_1x+a_2x^2+....$$

Then G (0) is simply equal to a_0, which is 0 in this case.

Hence, $G(x) = \ln(1+x)$.

4. Using generating function, let us solve:

$$y_{n+2}-5y_{n+1}+6y_n=0,\ n\ \geq\ 0 \text{ with } y_0=1 \text{ and } y_1=1.$$

Solution:

Given:

$$y_{n+2}-5y_{n+1}+6y_n=0 \qquad ...(1)$$

Multiply (1) by z^n and taking the sum over all n:

$$\sum_{n=0}^{\infty}y_{n+2}\ z^n-5\sum_{n=0}^{\infty}y_{n+1}\ z^n+6\sum_{n=0}^{\infty}y_n\ z^n=0 \qquad ...(2)$$

We know that:

$$G(y,z)=y_0+y_1z+y_2z^2+....+y_nz^n+....\infty$$

$$=\sum_{n=0}^{\infty}y_n\ z^n \qquad ...(3)$$

From(2)

$$\left[y_2+y_3z+y_4z^2+...\right]-5\left[y_1+y_2z+y_3z^2+...\right]+6G(y,z)=0$$

$$\frac{1}{z^2}\left[y_0z^2+y_3z^3+....\right]-\frac{5}{z}\left[y_1z+y_2z^2+...\right]+6G(y,z)=0$$

$$\frac{1}{z^2}\left[G(y,z)-y_0-y_1z\right]-\frac{5}{z}\left[G(y,z)-y_0\right]+6G(y,z)=0$$

Given $y_0 = 1$ and $y_1 = 1$.

$$z^{\bar{2}}\left[G(y,z)-1-z\right]-\frac{5}{z}\left[G(y,z)-1\right]+6G(y,z)=0.$$

Multiply by z²:

$$[G(y,\ z)-1-\ z]-\ 5z[G(y,\ z)-1]+\ 6\ z^2G(y,\ z)=0.$$

$$\text{i.e.,}(6\ z^2-\ 5z\ +1)G(y,\ z)-1-\ z\ +\ 5z\ =0$$

$$G(y,z)=\frac{1-4z}{6z^2-5z+1}$$

$$=\frac{(1-4z)}{(1-2z)(1-3z)} \qquad ...(4)$$

$$G(y,z)=\frac{2}{1-2z}-\frac{1}{1-3z} \qquad ...(5)$$

By partial fraction.

We know that:

$y_n = b(a)^n$ When $G(y,\ z)=\dfrac{b}{1-az}= 2(2)^n$ - $(3)^n$ is the solution of recurrence relation.

2.2 Recurrence Relations and Recursive Algorithms

The Fibonacci sequence is perhaps the most famous example of a recurrence relation or a definition made using recursion. Many concepts in mathematics and computer science need to be defined recursively.

Complete recursive definition has two parts:

Recursive Definition

A recursive definition of a concept is a definition that has two parts:

- Base: A definition of the concept for some starting point, usually n = 0 or n = 1 term.
- Recursion: How to get to the next instance of the concept based on previous instances or smaller choices of n.

The concept is related to the concept of induction. In general, we use recursion to define the sequences and we use induction to prove claims about recursively defined sequences.

Recurrence Relation (Epp)

A recurrence relation for a sequence a_0, a_1, a_2, ... is a formula that relates each term a_k to certain of its predecessors a_{k-1}, ..., a_{k-i}, where i is fixed and $k \geq i$. The initial conditions specify the fixed values of a_0, ..., a_{i-1}.

Most of the time, though, there is only one fixed value or base value. However, nothing prevents us from defining a sequence with multiple base values, hence the generality with i in the definition above.

Closed form of the Recurrence

It is an algebraic formula or a definition that tells us how to find the nth term without knowing any of the preceding terms. The process of finding the closed form is called solving a recurrence. There are various methods in “solving” recurrence that are used in practice.

Three Methods in Solving Recurrences

Iteration

Start with the recurrence and keep applying the recurrence equation until we get a pattern. The result is a guess at the closed form.

Substitution

Guess the solution and prove it using induction. The result here is a proven closed form. It is often difficult to come up with the guess so, in practice, iteration and substitution are used hand-in-hand.

Master Theorem

Plugging into a formula that gives an approximate bound on the solution. The result here is only a bound on the closed form. It is not an exact solution.

First-order Logic

The collection of terms of first-order logic or the first-order predicate calculus is defined by the following rules:

- A variable is a term.
- If f is an n-place function symbol (with $n \geq 0$) and $t_1,\ldots,t_n$ are terms, then $f(t_1,\ldots,t_n)$ is a term.

If P is an n-place predicate symbol and $t_1,\ldots,t_n$ are terms, then $p(t_1,\ldots,t_n)$ is an atomic statement.

Let us consider the sentential formulas $\forall$ x B and $\exists$ x B, where B is a sentential formula, $\forall$ is the universal quantifier ("for all") and B is the existential quantifier ("there exists").

B is called the scope of the respective quantifier and any occurrence of variable x in the scope of a quantifier is bound by the closest $\forall x$ or $\exists x$. The variable X is free in the formula B if at least one of its occurrences in B is not bound by any quantifier within B.

The collection of sentential formulas of first-order predicate calculus is defined by the following rules:

- Any atomic statement may be a sentential formula.
- If B and C are sentential formulas, then B (NOT B), B$\wedge$C(B AND C), B$\vee$C (B OR C) and B$\Rightarrow$ C (B implies C) are sentential formulas.
- If B is a sentential formula in which x is a free variable, then $\forall$ x B and $\exists$ x B are sentential formulas.

First-order Predicate Calculus

In the mathematical expressions of first-order predicate calculus, all the variables are object variables serving as arguments of functions and predicates. In second-order predicate calculus, variables may denote predicates and quantifiers may apply to variables standing for predicates.

The group of axiom schemata of first-order predicate calculus is comprised of the axiom schemata of propositional calculus together with the two following axiom schemata:

$$\forall x F(x) \Rightarrow F(r) \quad \ldots(1)$$

$$F(r) \Rightarrow \exists x F(x) \qquad ...(2)$$

Where,

F(x) is any sentential formula in which x occurs free.

r is a term.

F(r) is the result of substituting r for the free occurrences of x in sentential formula F and all occurrences of all variables in r are free in F.

Rules of inference in first-order predicate calculus are the Modus Ponens and the two following rules are as follows:

$$\frac{F(x)G \Rightarrow F(x)}{G \Rightarrow \forall x F(X)} \qquad ...(3)$$

$$\frac{F(X) \Rightarrow G}{\exists X F(X) \Rightarrow G} \qquad ...(4)$$

F(x) is any sentential formula in which x occurs as a free variable.

x does not occur as a free variable in formula G and the notation means that if the formula above the line is a theorem formally deducted from axioms by application of inference rules, then the sentential formula below the line is also a formal theorem.

Likewise to propositional calculus, rules for introduction and elimination of ∀ and ∃ can be derived in first-order predicate calculus.

For example, the following rule holds provided that F(r) is the result of substituting variable r for the free occurrences of x in sentential formula F and all occurrences of r resulting from this substitution are free in F:

$$\frac{\forall x F(X)}{F(r)}$$

Second Order Logic

The language of arithmetic is important to the foundation of mathematics. As axioms, we are able to take the general Pea no postulates, including even the second-order induction axiom.

In the standard semantics, the only model of the Pea no postulates, up to isomorphism, is the usual model of arithmetic. So, the theory generated by these axioms in the standard semantics is simply the second-order theory of true arithmetic.

The Pea no postulates have general models that may differ from the usual model in either or both of two ways. We can do the compactness theorem to construct the non-standard general models of the Pea no postulates containing infinitely large numbers.

We can also find general models of the Pea no postulates in which the universe of sets is less than the full power set of the individual universe. Indeed, any countable general model must be of this kind.

In the context of general models, we add the additional axiom schema for choice:

$\forall n\ \exists X\ \phi(n,X) \rightarrow \exists Y\ \forall n\ \phi(n,\{t \mid Ynt\})$ Prefixed by universal quantifiers as needed.

Here the formula that has been written as φ(n, {t | Ytn}) is obtained from φ(n,X), by replacing each term Xu by the term Ynu.

The Pea no postulates are strong enough to provide us with pairing functions. Consequently, for a general model, its 1-place relation universe completely determines its k-place relation universe for each k. The traditional terminology refers to the second-order number theory as analysis. The name derives from the fact, that it is possible to identify real numbers with sets of natural numbers.

The second-order quantifiers over sets of natural numbers can then be viewed as quantifiers over the real numbers. The appropriateness of the name is open to question, but its usage is well established. Accordingly, by a model of analysis we will mean a general model of the Pea no postulates with choice.

As a general model, any model of analysis must course fulfill all of the comprehension axioms.

Let A2 be the theory generated by the Pea no postulates with choice. Then A2 is a complete enumerable subset of true second-order arithmetic. A2 contains every true Σ-0-1 sentence. It does not contain every true Π-0-1 sentence.

We can obtain a stronger theory by restricting attention to the models of analysis that differ from the usual model. i.e., define an ω-model of analysis to be a model of analysis in which the individual universe is the actual set of natural numbers and the symbols 0 and S have their usual interpretations. Consequently, the symbols <, + and × have their usual interpretations.

The set universe of ω-model of analysis should be the power set of the natural numbers, but it must satisfy the full comprehension condition.

Problems

1. A sequence $\{a_n\}$ is defined recursively by $a_1 = 1$ and $a_n = 3a_{n-1} + 1$, for $n > 1$. Let us determine the first five terms of the sequence. Also form a recurrence relation of a_n and investigate that it is correct.

Solution:

Given:

$$a_1 = 1 \text{ and } a_n = 3a_{n-1} + 1, \text{ for } n > 1$$

Using the defined sequence, the first five terms an created like:

$$a_1 = 1$$

$$a_2 = 3a_1 + 1 = 3.1 + 1 = 4$$

$$a_3 = 3a_2 + 1 = 3.\ 4 + 1 = 13$$

$$a_4 = 3a_3 + 1 = 3.13 + 1 = 40$$

$$a_5 = 3a_4 + 1 = 3.40 + 1 = 121$$

To find a recurrence relation of an.

Since, a_1 is prescribed, let us start with a_2, i.e.,

$$a_2 = 31 + 1$$

$$a_3 = 32 + 3 + 1$$

$$a_4 = 33 + 32 + 31 + 1$$

$$a_5 = 34 + 33 + 32 + 31 + 1$$

Adding all and extending to nth term, we obtain:

$$a_n = 3^{n-1} + 3^{n-2} + ... + 3^2 + 3^1 + 1 = \frac{3^n - 1}{3-1} = \frac{3^n - 1}{2}$$

Which is to be investigated for its validity.

By principle of mathematical induction, we can prove that the above recurrence relation of an is justified and correct.

Assume n=1. Then, $a_1 = \frac{3^1 - 2}{2} = 1$. So, $a_n = \frac{3^n - 1}{2}$ is valid for n= 1.

Now, suppose $a_n = \frac{3^n - 1}{2}$ is valid for all positive integers n = 1,2,..., k, where k ≥1 and $a_k = \frac{3^k - 1}{2}$.

Then, $a_{k+1} = 3a_k + 1 = 3.\frac{3^k - 1}{2} + 1 = \frac{3^{k+1} - 3 + 2}{2} = \frac{3^{k+1} - 1}{2}$

Which is valid form, n = k + 1.

Thus, by the principle of mathematical induction, $a_n = \frac{3^n - 1}{2}$ is valid for all positive integers n = 1, 2,...., k. Finally, we may say that the newly formed recurrence relation, according to the given sequence, is valid and correct.

2.2.1 Linear Recurrence Relations with Constant Coefficients

1. Let us solve $a_r - 3a_{r-1} = 2$, r≥1 with ao =1 using the generating functions.

Solution:

Given:

$$a_r - 3a_{r-1} = 2$$

Multiplying both sides by z_r, we obtain:

$$a_r z^r - 3a_{r-1} z^r = 2z^r$$

Since r ≥1, summing for all r, we get:

$$\sum_{r=1}^{\infty} a_r z^r - 3a_{r-1} \sum_{r=1}^{\infty} z^r = 2\sum_{r=1}^{\infty} z^r$$

$$A(z) = a_0 + a_1 z + a_2 z^2 + ...$$

$$\sum_{r=1}^{\infty} a_r z^r = A(z) - a_0$$

$$\text{for} \sum_{r=1}^{\infty} a_{r-1} z^r = z\sum_{r=1}^{\infty} a_{r-1} z^{r-1}$$

$$= z\sum_{r=0}^{\infty} a_r z^r \quad [\text{replacing } r-1 \text{ by } r]$$

$$= zA(z)$$

$$\text{Also} \sum_{r=1}^{\infty} z^r = z + z^2 + z^3 + ... + z^r + ...$$

Using the sum of infinite terms of geometric progression (G.P),

$$S_n = a + ar + ar^2 + ...$$

Then, $S_\infty = \frac{a}{1-r}, r<1$

$$[A(z)-a_0]-3zA(z) = \frac{2z}{1-z}$$

or, $(1-3z)\,A(z) = \frac{2z}{1-z} + a_0$

$$a_0 = 1 \text{ given}$$

Hence, $A(z) = \frac{2z}{(1-z)(1-3z)} + \frac{1}{1-3z}$

or, $A(z) = \frac{1+z}{(1-z)(1-3z)}$

$$A(z) = \frac{2}{1-3z} - \frac{1}{(1-z)}$$

$$= 2(1-3z)^{-1} - (1-z)^{-1}$$

$$= 2[1+(3z)+(3z)^2+(3z)^3+......]-[1+z+z^2+z^3+....]$$

$$A(z) = 2\sum_{r=0}^{\infty}(3)^r z^r - \sum_{r=0}^{\infty}(1)^r z^r \quad \text{[Using binomial theorem]}$$

Hence,

$$a_r = 2(3)^r - (1)^r,\ r \geq 0$$

Which is the solution of the given recurrence relation.

2. Let us solve the recurrence relation:

$$a_n + 6a_{n-1} + 12a_{n-2} + 8a_{n-3} = 2^n,\ n \geq 3$$

$$a_0 = 0, a_1 = 0, a_2 = 2$$

Solution:

Given:

$$a_n + 6a_{n-1} + 12a_{n-2} + 8a_{n-3} = 2^n,\ n \geq 3$$

$$a_0 = 0, a_1 = 0, a_2 = 2$$

The characteristic equation is:

$$\lambda^3 + 6\lambda^2 + 12\lambda + 8 = 0$$

$$(\lambda + 2)^3 = 0$$

Therefore, $\lambda = -2, -2, -2$

Hence, the homogeneous solution is given by:

$$a_n^{(h)} = (A_1 n^2 + A_2 n + A_3)(-2)^n$$

The particular solution will be of the form P(2ⁿ). Substituting the given recurrence relation, we get:

$$P(2^n) + 6\ P(2^{n-1}) + 12\ P\ (2^{n-2}) + 8\ P\ (2^{n-3}) = 2n$$

$$2^n P\left[1 + \frac{6}{2} + \frac{12}{4} + \frac{8}{8}\right] = 2^n$$

Or,

$$P[1 + 3 + 3 + 1] = 1$$

$$P = \frac{1}{8}$$

$$a_n^{(P)} = \frac{1}{8}(2^n) = 2^{n-3}$$

Total solution is expressed as:

$$a_n = a_n^{(h)} + a_n^{(P)}$$

$$= (A_1 n^2 + A_2 n + A_3)(-2)^n + 2^{n-3}$$

Using the initial conditions:

$$a_0 = 0,\ a_1 = 0,\ a_2 = 2$$

$$0 = A_3 + \frac{1}{8}$$

$$A_3 = -\frac{1}{8} \qquad ...(1)$$

$$0 = (A_1 + A_2 + A_3)(-2)^1 + 2^{1-3}$$

$$A_1 + A_2 + A_3 = \frac{1}{8} \qquad ...(2)$$

Also $\quad 2 = (A_1 4 + A_2 \cdot 2 + A_3)(-2)^2 + 2^{2-3}$

$$2 = (4A_1 + 2A_2 + A_3)(4) + \frac{1}{2}$$

$$\frac{3}{2} = 4(4A_1 + 2A_2 + A_3)$$

$$4A_1 + 2A_2 + A_3 = \frac{3}{8} \qquad ...(3)$$

From equations (1), (2) and (3) $A_1 = 0,\ \ A_2 = \frac{1}{4},\ A_3 = -\frac{1}{8}$

$$a_n = \left(\frac{1}{4}n - \frac{1}{8}\right)(-2)^n + 2^{n-3}$$

2.2.2 Solution of Recurrence Relations by the Method of Generating Functions

We can determine the generating function of the sequence from the recurrence relation. One of the uses of generating function method is to find the closed form formula for a recurrence relation.

Once the generating function is known, an expression for the value of sequence may easily be obtained. Before using this method, ensure that the given recurrence equation is in linear form. A non-linear recurrence equation cannot be solved by the generating method. We use substitution of variable technique to convert a non-linear recurrence relation into linear equation.

Problems

1. Let us solve the recurrence relation $a_r - 3a_{r-1} + 2a_{r-2} = 0$ and r ≥2, by the generating function with the initial conditions $a_0 = 2$ and $a_1 = 3$.

Solution:

Given:

$$a_r - 3a_{r-1} + 2a_{r-2} = 0$$

$$r \geq 2$$

Initial conditions $a_0 = 2$ and $a_1 = 3$

Let A (Z) be the generating function of the sequence $<a_n>$ that is:

$$A(z) = \sum_{r=0}^{x} a_r Z'$$

Multiply the given recurrence relation by Z', we get:

$$a_r Z' - 3a_{r-1} Z' + 2a_{r-2} Z' = 0$$

Summing from r = 2 to x, we obtain:

$$\sum_{r-2}^{\infty} a_r Z' - 3\sum_{r-2}^{\infty} a_{r-1} Z' + 2\sum_{r-2}^{\infty} a_{r-2} Z' = 0$$

$$(A(Z) - a_0 - a_1 z) - 3z(A(Z) - a_0) + 2Z^2 A(Z) = 0$$

$$(2Z^2 - 3Z + 1)\ A(Z) - a_0 - a_1 Z + 3a_0 Z = 0$$

Now using the given conditions, i.e., $a_0 = 2$, $a_1 = 3$, we get:

$$(2\ Z^2 - 3Z + 1)\ A(Z) - 2 - 3Z + 6Z = 0$$

$$A(Z) = \frac{2 - 3Z}{2Z^2 - 3Z + 1}$$

$$= \frac{1}{(1-Z)} + \frac{1}{(1-2Z)}$$

Thus, $a_r = 1 + 2^r$.

2. Let us solve the recurrence relation $a_r - 2a_{r-1} + a_{r-2} = 2^r$, r≥2 by the generating function method with the boundary conditions $a_0 = 2$ and $a_1 = 1$.

Solution:

Given:

$$a_r - 2a_{r-1} + a_{r-2} = 2^r,\ r \geq 2$$

Boundary conditions:

$$a_0 = 2,\ a_1 = 1$$

Let A (Z) = $\sum_{r=0}^{\infty} a_r Z^r$ be the generating function of the sequence $<a_n>$, n≥0.

Multiply by Z^r in given recurrence relation and summing for all r≥2, we obtain:

$$\sum_{r=2}^{\infty} a_r 2^r - 2\sum_{r=2}^{\infty} a_{r-1} Z^r + \sum_{r=2}^{\infty} a_{r-2} Z^r = \sum_{r=2}^{\infty} 2^r Z^r$$

$$\left[A(Z) - a_0 - a_1 Z\right] - 2Z\left[A(Z) - a_0\right] + Z^2 A(Z) = \frac{4Z^2}{1-2Z}$$

Now, using the boundary conditions i.e., $a_0 = 2$ and $a_1 = 1$, we get:

$$\left(Z^2 - 2Z + 1\right)A(Z) - 2 + 3Z = \frac{4Z^2}{1-2Z}$$

$$\left(Z^2 - 2Z + 1\right)A(Z) = \frac{4Z^2}{1-2Z} + 2 - 3Z$$

So,

$$A(Z) = \frac{10Z^2 - 7Z + 2}{(1-2Z)(1-Z)^2}$$

$$= \frac{3}{1-Z} - \frac{5}{(1-Z)^2} + \frac{4}{1-2Z}$$

Thus, $a_r = 3 - 5\left(r + 1\right) + 4.2_r\ r \geq 2.$

3. Let us solve the recurrence relation $a_r - 5a_{r-1} - 6a_{r-2} = 2$, r≥2 by the generating function method with the boundary condition ao = 1 and a1 = 1.

Solution:

Given:

$$a_r - 5a_{r-1} - 6a_{r-2} = 2,\ r \geq 2$$

Boundary condition $a_0 = 1$ and $a_1 = 1$

Let $A(Z) = \sum_{r=0}^{\infty} a_r Z^r$ be the generating function of the sequence $<a_n>$, $n \geq 0$.

Multiply by Z^r in the given recurrence relation and summing for all $r \geq 2$, we obtain:

$$\sum_{r=2}^{\infty} a_r Z^r - 5\sum_{r=2}^{\infty} a_{r-1} Z^r + 6\sum_{r=2}^{\infty} a_{r-2} Z^r = \sum_{r=2}^{\infty} 2Z^r$$

$$(A(Z) - a_0 - a_1 Z) - 5Z(A(Z) - a_0) + 6Z^2 A(Z) = 2[Z^2 + Z^3 + ...]$$

Using the boundary conditions $a_0 = 1$ and $a_1 = 1$, we get:

$$A(Z)\ [1 - 5Z + 6Z^2] - 1 + 4Z = \frac{2Z^2}{1-Z}$$

$$A(Z)\ (6Z^2 - 5Z + 1) = \frac{2Z^2}{1-Z} + 1 - 4Z$$

$$A(Z) = \frac{2Z^2}{(1-Z)(1-2Z)(1-3Z)} + \frac{1-4Z}{(1-2Z)(1-3Z)} = \frac{1}{1-Z}$$

Consequently, we have $a_r = 1$.

4. Let us solve the recurrence relation $a_r - 7a_{r-1} + 10a_{r-2} = 0$ by the method of generating functions with the initial conditions $a_0 = 3$ and $a_1 = 3$.

Solution:

Given:

$$a_r - 7a_{r-1} + 10a_{r-2} = 0$$

Initial conditions $a_0 = 3$ and $a_1 = 3$

Let $A(Z) = \sum_{r=0}^{\infty} a_r Z^r$ be the generating functions of the equation $<a_n>$, $n \geq 0$.

The given recurrence relation is:

$$a_r - 7a_{r-1} + 10a_{r-2} = 0 \qquad ...(1)$$

Multiply equation (1) by Z^r and summing from r = 2 to ∞, we get:

$$\sum_{r=2}^{\infty} a_r Z^r - 7\sum_{r=2}^{\infty} a_{r-1} Z^r + 10\sum_{r=2}^{\infty} a_{r-2} Z^r = 0$$

$$\left[A(Z) - a_0 - a_1 Z\right] - 7Z\left[A(Z) - a_0\right] + 10Z^2 A(Z) = 0$$

Now, using the initial conditions i.e., $a_0 = 3$ and $a_1 = 3$, and solving it, we get:

$$A(Z) = \frac{3+24Z}{10Z^2 - 7Z + 1} = \frac{3+24Z}{(5Z-1)(2Z-1)}$$

$$A(Z) = \frac{13}{(1-5Z)} - \frac{10}{(1-2Z)}$$

Hence, $a_r = 13(5)^r - 10\ (2)^r$.

5. Using generating function method, let us solve the recurrence relation $Y_{n+2} - 4y_{n+1} + 3y_n = 0$ with the initial conditions $y_0 = 2$, $y_1 = 4$.

Solution:

Given:

$$Y_{n+2} - 4y_{n+1} + 3y_n = 0$$

$$y_0 = 2,\ y_1 = 4$$

Let the generating function of the equation $<y_n>$, n≥0 be:

$$A(Z)\ \sum_{n\ 0}^{\infty} y\ Z \qquad ....(A)$$

The given recurrence equation is:

$$Y_{n+2} - 4Y_{n+1} + 3Y_n = 0 \qquad ...(1)$$

Multiply (1) by Z^n and summing from n = 2 to ∞, we get:

$$\sum_{n=2}^{\infty} y_{n+2} \cdot Z^n - 4\sum_{n=2}^{\infty} y_{n+1} Z^n + 3\sum_{n=2}^{\infty} y_n Z^n = 0$$

$$= \frac{1}{Z^2}\left[A(Z) y_0 - y_1 Z\right] - 4 \cdot \frac{1}{Z}\left[A(Z) - y_0\right] + 3A(Z) = 0$$

Now using the conditions i.e. $y_0=2$ and $y_1=4$, we get:

$$\frac{1}{Z^2}\left[A(Z)-2-4Z\right]-\frac{4}{Z}\left[A(Z)-2\right]+3A(Z)=0$$

$$A(Z)[1-4Z+3Z^2]=2-4Z$$

$$A(Z)=\frac{2-4Z}{1-4Z+3Z^2}=\frac{2-4Z}{(1-Z)(1-3Z)}$$

$$A(Z)=\frac{1}{(1-Z)}+\frac{1}{(1-3Z)}$$

Comparing this with equation (A), we get:

$$y_n=1+3n$$

2.3 Divide and Conquer Algorithms

Divide and conquer is a general algorithm design strategy with a general plan as follows:

Divide

A problem's instance is divided into several smaller instances of the same problem, ideally of about the same size.

Recur

Solve the sub-problem recursively.

Conquer

If necessary, the solutions obtained for the smaller instances are combined to get a solution to the original instance. The base case for the recursion is a sub-problem of constant size.

Advantages of Divide and Conquer Technique:

- For solving conceptually difficult problems like Tower Of Hanoi, divide & conquer is a powerful tool.
- Results in efficient algorithms.

- Divide & Conquer algorithms are adapted for execution in multi-processor machines.
- Results in algorithms that use memory cache efficiently.

Limitations of Divide and Conquer Technique:

- Recursion is slow.
- Very simple problem may be more complicated than an iterative approach. Example: Adding n numbers, etc.

Divide and conquer algorithm is easier to solve several small instances of a problem than one large one. They divide the problem into smaller instances of the same problem then solve the smaller instances recursively and finally combine the solutions to obtain the solution for the original input.

To design a divide and conquer algorithm, we must specify the subroutines that directly solve or divide and combine.

n - Size of the input.

k - Smaller instances of n.

B (n) - Number of steps done by directly solving.

D (n) - Number of steps done by dividing.

C (n) - Number of steps done by combining.

With the above specification, the final recurrence equation that describes the amount of work done by the algorithm is:

$$T(n) = D(n) + \sum_{i=1}^{k} T(\text{size}(I_i)) + C(n) \text{ for } n > \text{Small size.}$$

T (n) = B (n) for n ≤ Small size.

The skeleton procedure for divide and conquer is:

Solve (I)

n = size (I);

If (n ≤ small Size)

Solution = directly Solve (I);

Else

Divide I into $I_1 \ldots I_k$

For each i ∈ {1, ...k};

S_i = solve (I_i);

Solution = combine (S_1, ...S_k);

Return solution.

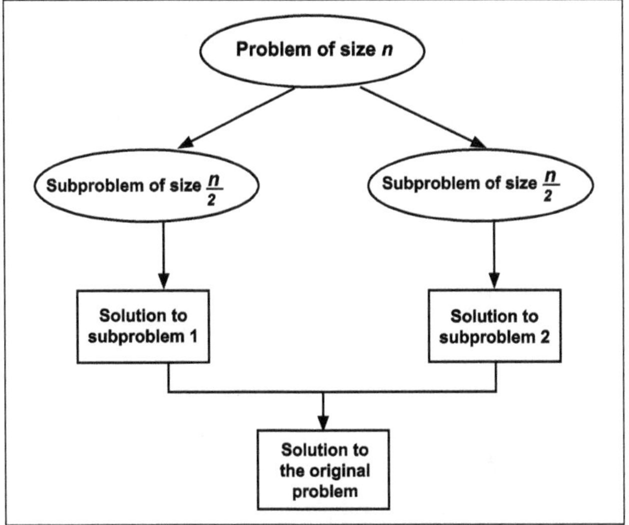

Divide and conquer technique.

Generally, an instance of size n can be divided into several instances of size n/b, with a number of times T(n/b) to be computed. That is the recurrence equation for the running time T(n) is given by,

$$T(n) = aT(n/b) + f(n)$$

a, b - Constants where $a \geq 1$ and $b > 1$

n - Power of b

f (n) - A function that accounts for the time spent on dividing the problem into smaller ones and combining their solutions.

The order of growth of T(n) depends on the values of the constants a, b and the function f(n).

Examples:

- Sorting: Merge sort and quicksort.
- Binary tree traversals.
- Binary search.
- Multiplication of large integers.

- Matrix multiplication: Stassen's algorithm.
- Closest pair and convex hull algorithms.

Master Theorem

$$T(n)= aT(n/b)+f(n) \text{ where } f(n)\in\Theta\left(n^{d}\right), d\geq 0$$

$$\text{If } a<b^{d}, T(n)\in\Theta\left(n^{d}\right)$$

$$\text{If } a=b^{d}, T(n)\in\Theta\left(n^{d}\log n\right)$$

$$\text{If } a>b^{d}, T(n)\in\Theta\left(n^{d}\log_{b}a\right)$$

The same results hold with O, instead of Θ.

Example:

Computing the sum of n numbers $a_0 ....a_{n-1.}$

If n = 1, a_0 is returned as answer.

If n > 1, then:

$$a_0+.....+a_{n-1}=\left(a_0+....+a_{\left\lceil\frac{n}{2}\right\rceil-1}\right)+\left(a_{\left\lceil\frac{n}{2}\right\rceil}+...+a_{n-1}\right)$$

The recurrence relation is:

$$A(n)=2A\left(\frac{n}{2}\right)+1$$

Where, a = 2, b = 2 and d = 0 (i.e.) $n^d = n^0=1$.

Here, a > bd(i.e.) 2 > 20.

$$\therefore\ A(n)\in\Theta(n\ \log_{b}a)$$

$$\in\Theta(n\ \log_{2}2)$$

$$\in\Theta\left(n^{1}\right)$$

$$\in\Theta(n).$$

3

Groups and Rings, and Boolean Algebras

3.1 Groups and Subgroups

Groups

A group is any set of objects with an associated operation that combines pairs of objects in the set. In other words, a group is defined as a set G together with a binary operation. We will use * (or sometimes 'o') to denote the operation although this does not imply that groups only apply to multiplication. An example of a group might be the set of all integers with the operation of addition.

Groups provide a level of abstraction apart from the mathematical notations. For example, rotations might be modeled by matrices or by quaternions or by multi vectors or by some other notation, however we may wish to study the properties of rotations without getting involved with the mechanics of matrices, etc.

Two properties of a group are as follows:

- The identity element of a group is unique.
- The inverse of each element is unique.

Dihedral Group (D_4, *) and its Composition Table

The set of transformation due to all rigid motions of a square resulting in identical squares but with different vertex names under the binary operation of right composition * is a group called dihedral group of order 8 and denoted by {D_4, *}.

By rigid motion, the rotation of the square about its centre through angles 90°, 180°, 270°, 360° in the anticlockwise direction and reflection of the square about 4 lines of symmetry is as given in the below figure:

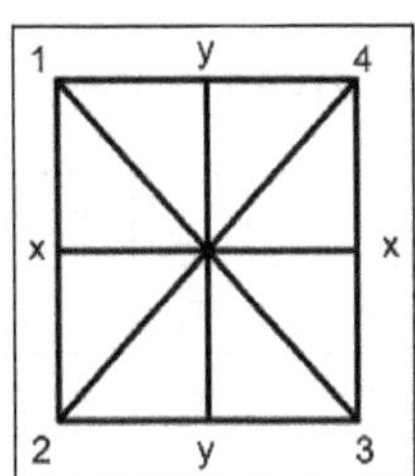

$$r_1 = r(90°) = \begin{pmatrix} 1 & 2 & 3 & 4 \\ 4 & 1 & 2 & 3 \end{pmatrix}$$

$$r_2 = r(180°) = \begin{pmatrix} 1 & 2 & 3 & 4 \\ 3 & 4 & 1 & 2 \end{pmatrix}$$

$$r_3 = r(270°) = \begin{pmatrix} 1 & 2 & 3 & 4 \\ 2 & 3 & 4 & 1 \end{pmatrix}$$

$$r_4 = r(360°) = \begin{pmatrix} 1 & 2 & 3 & 4 \\ 1 & 2 & 3 & 4 \end{pmatrix}$$

$$r_5 = r(XX) = \begin{pmatrix} 1 & 2 & 3 & 4 \\ 2 & 1 & 4 & 3 \end{pmatrix}$$

$$r_6 = r(YY) = \begin{pmatrix} 1 & 2 & 3 & 4 \\ 4 & 3 & 2 & 1 \end{pmatrix}$$

$$r_7 = r(13) = \begin{pmatrix} 1 & 2 & 3 & 4 \\ 1 & 4 & 3 & 2 \end{pmatrix}$$

$$r_8 = r(24) = \begin{pmatrix} 1 & 2 & 3 & 4 \\ 3 & 2 & 1 & 4 \end{pmatrix}$$

Now,

$$\begin{array}{lcccc} r_1 \bullet r_1 = r_1 & 1 & 2 & 3 & 4 \\ & \downarrow & \downarrow & \downarrow & \downarrow \\ & 4 & 1 & 2 & 3 \\ r_1 & \downarrow & \downarrow & \downarrow & \downarrow \\ & 3 & 4 & 1 & 2 \end{array}$$

$$= \begin{pmatrix} 1 & 2 & 3 & 4 \\ 3 & 4 & 1 & 2 \end{pmatrix} = r_2$$

Similarly, the composition table is as given below:

*	r_1	r_2	r_3	r_4	r_5	r_6	r_7	r_8
r_1	r_2	r_3	r_4	r_1	r_8	r_2	r_5	r_6
r_2	r_3	r_4	r_1	r_2	r_6	r_5	r_8	r_3

r_3	r_4	r_1	r_2	r_3	r_7	r_8	r_6	r_5
r_4	r_1	r_2	r_3	r_4	r_5	r_6	r_7	r_8
r_5	r_7	r_6	r_8	r_5	r_4	r_2	r_1	r_3
r_6	r_8	r_5	r_7	r_6	r_2	r_4	r_3	r_1
r_7	r_6	r_8	r_5	r_7	r_3	r_1	r_4	r_2
r_8	r_5	r_7	r_6	r_8	r_1	r_3	r_2	r_4

From the table, $r_4 * r_i = r_i * r_4 = r_i$; $I = \{1, 2, \dots, 8\}$.

$\therefore$ r_4 is the identity element of $\{D_4, *\}$.

Also from the table, we see that the inverses of r_1, r_2, r_3, r_4, r_5, r_6, r_7 and r_8 are respectively r_3, r_2, r_1, r_4, r_5, r_6, r_7, r_8.

Problems

1. If a and b are any two elements of a group (G, *), let us show that G is an Abelian group if and only if $(a*b)^2 = a^2*b^2$.

Solution:

Given:

$$(a*b)^2 = a^2*b^2$$

$\Rightarrow$ Part

G is an abelian; $a * b = b * a$ ...(1)

Now, $(a*b)^2 = (a*b)*(a*b)$

$$= a*(b*a)*b$$

$$= a*(a*b)*b$$

$$= (a*a)*(b*b), \text{ by associative law}$$

$$= a^2 * b^2$$

$\Rightarrow$ Part

$$(a*b)^2 = a^2 * b^2$$
$$(a*b)*(a*b) = (a*a)*(b*b)$$
$$a*(b*a)*b = a*(a*b)*b$$
$$(b*a) = a*b$$

By left and right cancellation law.

Here, (G, *) is abelian:

i.e., G is an abelian group if $f(a * b)^2 = a^2 * b^2$.

2. Let us show that any group G is abelian if $(ab)^2 = a^2b^2$ for all a, b ∈ G.

Solution:

Given:

$$(ab)^2 = a^2b^2 \text{ for all } a, b \in G \cdot$$

G is abelian group.

$$\therefore ab = ba; \text{ for all } a,\ b \in G \quad ...(1)$$

Now, $(ab)^2 = (ab)(ab)$

$$= a(ba)b, \text{ by associative law}$$

$$= a(ab)b \text{ by (1)}$$

$$= (aa)(bb)$$

$$(ab)^2 = a^2b^2.$$

Conversely, Let G be a group satisfying:

$(ab)^2 = a^2b^2$; for all a, b ∈ G.

To prove that it satisfies commutative property:

Given:

$$(ab)^2 = a^2b^2$$

$$\Rightarrow (ab)(ab) = (aa)(bb)$$

$$\Rightarrow a(ba)b = a(ab)b, \text{ by associative law}$$

$$\Rightarrow ba = ab, \text{ by left and right}$$

Cancellation law:

G is an abelian group.

3. Let us show that the inverse of an element in a group (G, *) is unique.

Solution:

Given:

Let (G, *) be a group with identity element e.

Let b and c be inverses of an element a ∈ G such that:

$$a*b=b*a=e \quad ...(1)$$

$$a*c=c*a=e \quad ...(2)$$

Now, $b=b*e=b*(a*c)$ by (2)

$=(b*a)*c$; associate law

$=e*c$ by (1)

$=c$

Hence, Inverse is unique.

Subgroups

1. If H_1 and H_2 are sub-groups of a group (G, *), then $H_1 \cap H_2$ is a sub-group of G.

H_1, H_2 be subgroups of (G, *) and e is the identity element of G. Now as $e \in H_1$ and $e \in H_2$; $e \in H_1 \cap H_2$

Hence, $H_1 \cap$ H2 is non-empty. ...(1)

Let a, b ∈ $H_1 \cap H_2$ then a, b ∈ $H_1$and a, b ∈ H_2. Since H_1 and H_2 are subgroups of G, a * b-1 ∈ H_1 and a * b-1 ∈ H_2.

∴ a * b-1 ∈ $H_1 \cap H_2$ ...(2)

From (1) & (2),

$H_1 \cap H_2$ is a subgroup of G.

2. The Kernel of a homomorphism f from a group (G, *) to (H, Δ) is a normal subgroup of G.

(G, *) to (H, Δ) is a normal subgroup of G.

A subgroup {H, *} of the group {G, *} is called a normal subgroup, if for any a ∈ G; aH = Ha.

Let f: G → H be a group homomorphism.

& ker $f = \{x \in G\ [f(x) = e]\}$ be the kernel of f.

"e" being the identity element of H.

As f(e) = e we have e ∈ ker f.

Let x, y ∈ ker f then:

$$f(x * y^{-1}) = f(x)\ \Delta\ f(y^{-1})$$
$$= f(x) \Delta (f(y))^{-1}$$
$$= e \in H$$

and hence $x * y^{-1} \in$ ker f.

Thus, ker f is a subgroup of G.

To prove that ker f is normal.

Let a ∈ G and x ∈ ker f. Then f(x) = e

And $f(a * x * a-1) = f(a)\ \Delta\ f(x)\ \Delta\ f(a\_1)$

$$= f(a) \Delta\ e\ \Delta (f(a))^{-1}$$
$$r = f(a) \Delta (f(a))^{-1} \text{ as } f(x) = e'$$
$$= e' \,\&\, f(a^{-1}) = (f(a))^{-1}.$$

Thus, $a\ x\ a^{-1} \in$ ker f for all a ∈ G and x ∈ kerf.

Hence, ker f is a normal subgroup of G.

3. Every subgroup of an abelian group is normal.

Let H be a subgroup of an abelian group G.

$$a,\ b \in G;\ ab = ba.$$

Let x ∈ H & a ∈ G.

Now, $ax\,a^{-1} = (xa)a-1$ (Since H is subgroup x ∈ G)

$$= x(aa^{-1})$$

$$= xe$$

$= x \in H$

$ax\,a^{-1} \in H$. Hence, H is normal.

3.1.1 Cosets and Lagrange's Theorem

Normal Subgroups and Cosets

A subgroup $H \triangleleft G$ is normal if g Hg^{-1} for all $g \in G$.

Notation: $H \triangleleft G$

Every subgroup of an abelian group is normal. Every subgroup of index 2 is normal.

If $H \triangleleft G$, the set G|H of left cosets becomes a group under coset multiplication: (aH) (bH) = (ab)H . In this case, G|H is the quotient group (or factor group) of G by H.

The kernel of a group map is a normal subgroup. The image of a normal subgroup is a normal subgroup, provided that the group map is surjective.

Lagrange's Theorem

If G is a finite group and H is a sub-group of G, then O(H) is a divisor of O(G).

(or)

The order of a subgroup of a finite group G divides the order of the group G.

Lagrange's Theorem for Finite Groups

Let H be a subgroup of a finite group. Then order of H divides the order of G.

Proof:

Since G is finite group, H is also a finite subgroup. Let Ha and Hb be two right cosets of H is G.

Claim:

$$Ha = Hb \text{ (or) } Ha \cap Hb = \{\ \}$$

Suppose that $Ha \wedge Hb \neq \{\ \}, x \in Ha \cap Hb$

$\Rightarrow x \in Ha$ and $x \in Hb$

$\Rightarrow x = h_1 a,\ x = h_2\ b$ for some $h_1, h_2 \in H$

$\Rightarrow h_1, a = h_2 b$ ...(1)

Let $y \in Ha$

$\Rightarrow y = ha$ for some $h \in H$

$= h\left[h_1^{-1}h_2\ b\right]$ using (1)

$=\left(hh_1^{-1}\ h_2\right)b$ where $hh^{-1}\ h_2 \in H$

$\therefore y \in Hb \Rightarrow Ha \subseteq Hb$ ...(2)

Conversely, Let $z \in Hb \Rightarrow z = hb$ for some $h \in H$.

Then, $z = h\left(h_1^{-1}\ a\right)$ (using (1))

$=\left(hh_2^{-1}h_1\right)a$, where $hh_2^{-1}h_1, \in H$

$\therefore\ z \in Ha$

$\therefore Hb \subseteq Ha$ ...(3)

From (2) and (3),

Finally, $Ha = Hb$

Therefore the claim is true, (i.e) any two right cosets of H in G are either identical (or) disjoint.

∴ H = Ha is a right co set of H in G. Further since G is a finite group, then G has only finite number 'K' of right cosets in which distinct right co sets are disjoint. Since there is one to one correspondence between any two right cosets of H in G, each right coset has O(H) elements.

We know that a ∈ G is in the unique right coset Ha. Then:

$$G = \bigcup_{a \in G\cdot} Ha.$$

$$\Rightarrow O(G) = \sum_{a \in G} O(H).$$

Where this sum runs over each elt from every different co set of H in G.

$$\therefore O(G) = K\, O(H)$$

∴ O (H) divides O (G) and k is called index of H in G (or) the number of right cosets of H in G.

3.2 Codes and Group Codes

The binary code is the process of assigning the bit patterns to discrete elements of information. There can be different ways of assigning the bit patterns to this information, one way could be the straight forward assigning of binary codes which are binary equivalent of their numeric value. Another approach could be assigning some other combination of bits.

Table: Two different ways of assigning binary codes.

Decimal code	Binary code format - 1	Binary code format - 2
0	0000	0000
1	0001	0010
2	0010	0100
3	0011	0110
4	0100	1000
5	0101	1001
6	0110	0111
7	0111	0101
8	1000	0011
9	1001	0001

The relation between the number of bits in code word and the number of unique code words is, N = 2j

Where,

j is the number of bits.

N is the total number of code words.

The binary code can be of following types:

- Weighted Binary Codes.
    - The Weighted Binary Codes are of 8421 or BCD code, 2421 code etc.
- Non-weighted binary code.
    - Non-weighted binary code is gray code.

Binary Coded Decimal Codes

BCD codes are the weighted binary codes. The other examples of weighted codes are 2421, 5421, 5411 code. These codes obey their positional weighting principles. The bits are multiplied by their weights the sum of these weighted binary bits gives the equivalent decimal digit in the weighted binary system. All the BCD codes are used to represent the decimal numbers. Some other examples of BCD codes are excess-3 Code (XS-3), 84-2-1 code (+8, +4, -2, -1), Biquinary code, 7421 code, 5311 code etc.

Table: Example of BCD codes.

Decimal	XS-3	8421	8421	Biquinary ABCDEFG	2421	5311	5421
0	0011	0000	0000	0100001	0000	0000	0000
1	0100	0001	0111	0100010	0001	0001	0001
2	0101	0010	0110	0100100	0010	0011	0010
3	0110	0011	0101	0101000	0011	0100	0011
4	0111	0100	0100	0110000	0100	0101	0100
5	1000	0101	1011	1000001	1000	0111	1000
6	1001	0110	1010	1000010	1001	1001	1001
7	1010	0111	1001	1000100	1010	1011	1010
8	1011	1000	1000	1001000	1011	1100	1011
9	1100	1001	1111	1010000	1100	1101	1100

The BCD codes have certain properties which make it very distinctive and thus have the advantage over the other code. These properties are:

- Self-complementing property.
- Reflective property.

Self-complementing Codes

In the BCD system, some code have self-complementing property. i.e., Their logical complement is the same as its arithmetic equivalent. If the 9's complement of XS-3 is the same as its logical complement helps to find the extensive use in decimal arithmetic.

Examples:

6311 code, 2421 code, XS-3 code.

We can verify the self-complementing property of these codes from the above table.

Reflective Codes

This code is reflected about the center entries with one bit changed. It is said that the 9's complement of a reflected BCD code word is formed simply by changing only one of its bit.

Examples:

2421 code, 5211 code and XS-3 code.

Advantages of BCD Codes

- It is similar to the decimal system.
- The binary equivalent of decimal numbers 0 to 9 only.

Disadvantages of BCD Codes

- The BCD arithmetic is little more complicated.
- The addition and subtraction of BCD have different rules.
- BCD needs more number of bits than the binary for representing the decimal number. So, the BCD is less efficient than binary.

Group Codes

An (m, n) encoding function e: $B^m \to B^n$ is called a group code if $e(B^m) = \{e(b):\ b \in B^m\}$ is a sub-group of the group $(B^n,\oplus)$.

Since, $(B^n,\oplus)$ is an abelian group, the subgroup $e(B^m)$ shall be a normal subgroup of $(B^n, \oplus)$, We therefore represent $e(B^m)$ by N.

Problem

Show that the encoding function e: $B^2 \to B^5$ defined by

e(00)= 00000, e(01)= 01110, e(10)= 10101, e(11)= 11011 is a group code.

Solution:

Given:

$e(B^2)$= N ={00000,01110,10101,11011).

The addition table for N is

⊕	00000	01110	10101	11011
00000	00000	01110	10101	11011
01110	01110	00000	11011	10101
10101	10101	11011	00000	01110
11011	11011	10101	01110	00000

We note that for every x, y$\in$ N, x ⊕ y $\in$N. Therefore ⊕ is binary operation. The element 00000 acts as identity for N. Also, x⊕ x = 00000 for every x $\in$N showing that every element is inverse of itself. Hence, N is a subgroup of B5 and so the encoding function e is a group code.

3.2.1 Error Detection and Correction using Group Codes

Error Detecting Codes

When the digital information in the binary form is transmitted from one circuit or system to another circuit or system, an error may occur. To maintain the data integrity between transmitter and receiver, the extra bit or more than one bit are added in the data. The data along with the extra bit/bits forms the code.

Codes which allow only error detection are called error detecting codes. This means a signal corresponding to 0 may change to 1 or vice-versa due to presence of noise. To maintain the data integrity between transmitter and receiver, extra bit or more than one bit is added in the data. These extra bits allow the detection and some times, correction of error in the data.

The data along with the extra bit/bits forms the code. Codes which allow only error detection are called error detecting codes and codes which allow error detection and correction are called error detecting and correcting codes.

The most common cause for errors are that the noise creep into the bit stream during the course of transmission from transmitter to the receiver. If these errors are not detected and corrected, the result could be worst as the digital systems are too much sensitive to errors and will malfunction due to the slightest of errors in transmitted codes.

The various methods of error detection and correction such as addition of extra bits that is called check bits, sometimes they are also called redundant bits as they don't have any information in them. The various codes used for error detection and correction code in digital system are:

- Two-dimensional Parity check.
- Simple Parity check.
- Checksum.
- Cyclic redundancy check.

Error Correcting Codes

The method that we have discussed so far can detect errors but it can't correct them. Error Correction can be handled in 2 ways:

- Once when an error is discovered, the receiver might have the sender to retransmit the entire data unit. This is termed as backward error correction.
- The receiver can use an error-correcting code which automatically corrects certain errors. This is known as forward error correction.

In the theory, it is possible to correct any number of errors automatically. The error-correcting codes are more sophisticated than the error detecting codes and need more redundant bits. The number of bits required correcting the multiple-bit or burst error is so high that in most of the cases it is inefficient to do so. For this same reason, most error correction is limited to one, 2 or at the most three-bit errors.

Single-bit Error Correction

The concept of error-correction can be easily understood by examining the simplest case of single-bit errors. A single-bit error shall be detected by addition of a parity bit (VRC) with the data which required being send. A single additional bit can detect error, but it is not enough to correct that error too. For collecting an error one has to know the exact position of error, i.e. exactly which bit is in error.

For example, to correct the single-bit error in an ASCII character, the error correction must determine which one of the seven bits is in error. To this we have to add some additional redundant bits. For calculating the numbers of redundant bits (r) required to correct (d) data bits, let us determine the relationship between the two. Therefore, we have (d+ r) as the total number of bits which are to be transmitted, then r should be able to indicate at least d+r+1 different value.

Of these, one value means no error and remaining d +r values indicate error location of the error in each of d +r locations. So. d+ r+ l states must be distinguishable by r bits and r bits can indicates 2r states. Thus, 2^r must be greater than d+r+1,i. e. 2r > d+r+1. The value of r must be determined by putting in the value of d in the relation.

Example:

If d is 7, the smallest value of r that satisfies the above relation is 4. Thus, the total bits which are to be transmitted is 11 bits (d + r =7+4 =11).

Parity Checking

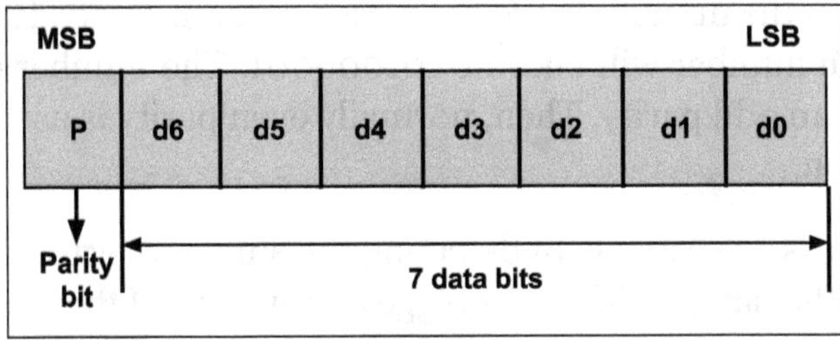

Parity Code

Parity Code is the simplest technique which is used for detecting and correcting errors. The MSB of an 8-bits word is used as the parity bit and the remaining 7 bits are used for data or message bits. Parity bit is added to the transmitted string of bits during the transmission from the transmitters to detect any error in the data when they are received at the receiver end. It is an extra bit added to the string of data. The parity of 8-bits transmitted word can be either even parity or odd parity.

Parity checking at the receiver can detect the presence of an error if the parity of receiver signal is different from the expected parity. That means, if it is known that the parity of the transmitted signal is always even and if the received signal has an odd parity, then the receiver can conclude that the received signal is not correct. When an error is detected, then the receiver will ignore the received byte and request for the retransmission of the same byte to the transmitter.

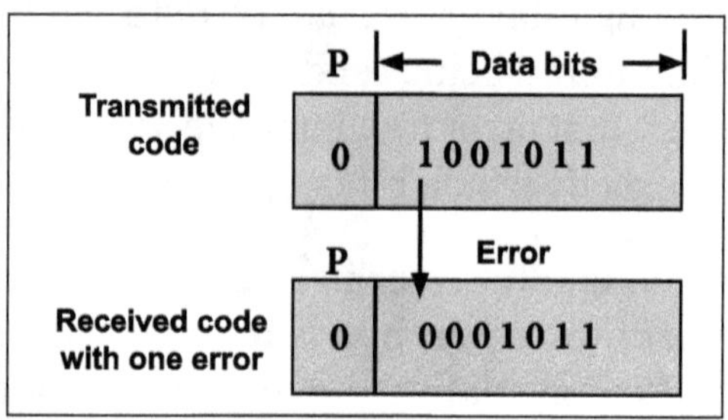

Even Parity

Even parity means the number of 1's in the given word including the parity bit should be even. When we get an even parity, after adding that extra bit the total number of 1's in the string of the data is even.

Odd Parity

Odd parity is defined as the number of 1's in the given word including that the parity bit should be odd. When we get an odd parity, after adding that extra bit into the data string the total number of 1's in the data shall be odd. We can understand it with an example, when we have an eight bit ASCII code - 01000001.

Now if the added bit is 0, then the number will become 001000001. In odd parity, the total number of 1's in the number is even so we get an even parity. Repeatedly if we add 1 to the number the number will become 101000001. The number of 1's is 3 which is odd so we have got an odd parity. Then, normally even parity is used and it has almost become a convention.

Now the parity checks are capable of detecting the single bit error. Since it fails, there are two changes in the data, which is the biggest drawback of this system. Hence, there are several other codes to detect and correct more than one bit errors.

Uses of Parity Bit:

- The parity bit can be set to 0 and 1 where it depends on the type of the parity required.

- For even parity, this bit is set to 1 or 0 whereas the no. of the 1 bits in the entire word is even which is shown in figure (a).
- For odd parity, this bit is set to 1 or 0 whereas the no. of the 1 bits in the entire word is odd which is shown in figure (b).

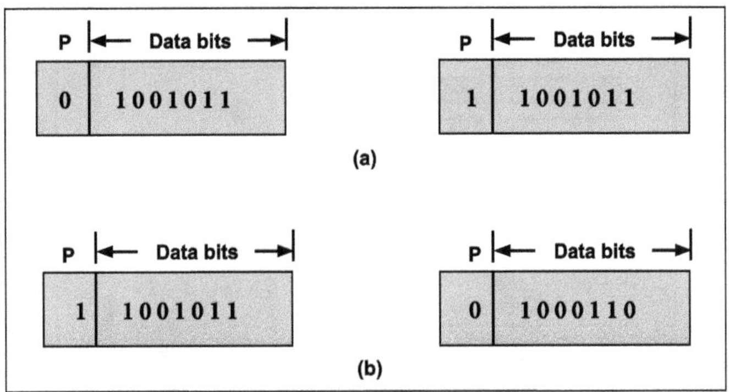

(a) Even parity and (b) Odd parity.

Hamming Code

Let us examine the manipulation with these bits to discover which bit is in error. The method developed by R.W. Hamming gives a practical solution. The solution or coding scheme, he developed is commonly called as Hamming Code. The hamming code can be applied to data units of any length and which uses the relationship between the data bits and the redundant bits.

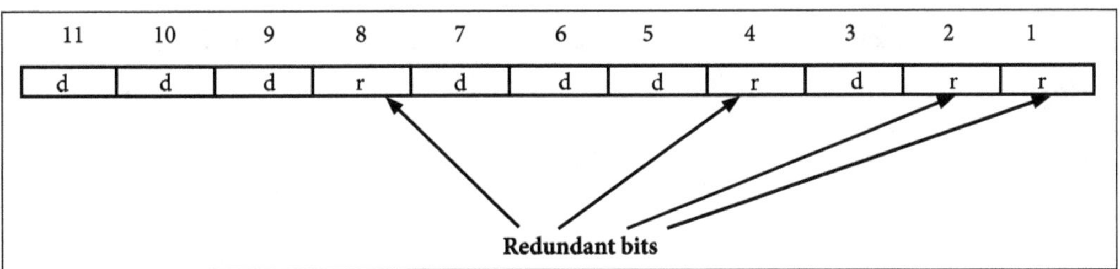

Positions of redundancy bits in hamming code.

The basic method for error detection by using Hamming code is as follows:

- The location of each of the (m+K) digits is assigned a decimal value.
- To each group of m information bits k parity bits are added to form (m+K) bit code as represented in figure.
- At the receiving end, the parity bits can be recalculated. The decimal value of the k parity bits provides the bit-position in error, if any.
- The k parity bits are placed in positions 1, 2, 2K-1 and K, parity checks are performed on selected digits of each code word.

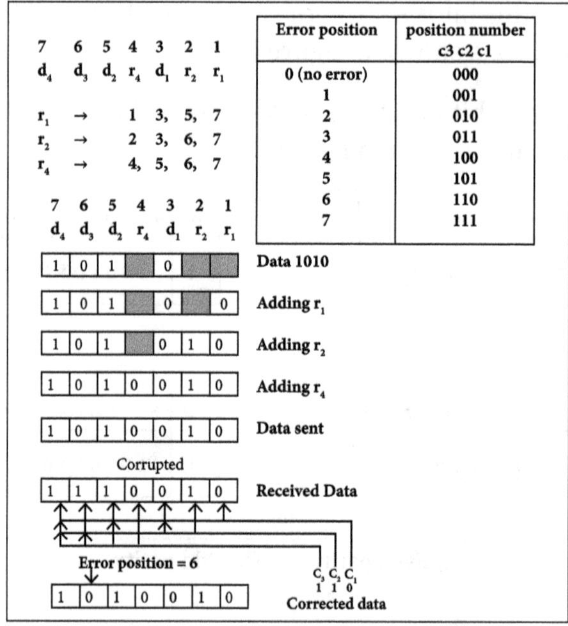

Error position	position number c3 c2 c1
0 (no error)	000
1	001
2	010
3	011
4	100
5	101
6	110
7	111

Use of hamming code for error collection for a 4-bit data.

The above figure represents how the hamming code is used for the correction of 4-bit numbers ($d_4d_3d_2d_1$) with the help of three redundant bits ($r_3r_2r_1$). For example, consider data 1010. First r_1 (0) is determined considering the parity of the bit positions 1, 3, 5 and 7. Then the parity bits r_2 is calculated by considering bit positions 2, 3, 6 and 7. Finally, the parity bits r_4 is calculated by considering the bit positions 4, 5, 6 and 7 as shown.

If any corruption occurs in any of the transmitted code 1010010, the bit position in the error can be found out by calculating the $r_3r_2r_1$ at the receiving end. For example, if the received code word is 1110010, the recalculated value of the r3r2r1 is 110 that indicate that bit position in error is 6, the decimal value of 110.

Parity Generator

It is the circuit that generates the parity bit in the transmitter.

Parity Checker

It is the circuit that checks the parity in the receiver.

3.3 Isomorphism

To show that two groups are isomorphic, an isomorphism from one to the other must be found. To show that two groups are not isomorphic, a group-theoretic property must be found which one group has but the other does not.

Given two groups (G1, *) and (G2, °) a useful first step in the search for an isomorphism is to write down the orders of the elements of each.

At this stage, we can see whether a objective function with order-preserving properties is possible. If it is not, then we can deduce immediately that the groups are not isomorphic. If it is possible to define such a function, then in our search for an isomorphism, some bijections can be eliminated by the fact that the orders of each element and its image must be equal.

If an order-preserving bijective function exists, this is not sufficient for us to conclude that the groups are isomorphic, since the property f (x*y) = f(x) o f(y) must also be satisfied. However, order-preserving bijections are the only possible candidates for isomorphisms.

Problems

1. Let us prove $(G_1 \text{ x } G_2)/(K_1 \text{ x } K_2)$ is isomorphic to $G_1/K_1 \text{ x } G_2/K_2$.

Solution:

Given:

First, we need to show a homomorphism exists, such that:

$$f: G_1 \times G_2 \to G_1/K_1 \times G_2/K_2.$$

Second, we need to verify:

$$\text{Im } f = G_1/K_1 \times G_2/K_2.$$

Last, we need to verify:

$$\text{Ker } f = K_1 \times K_2.$$

Then, we have defined a homomorphism such that:

Where g_1, g_2 are elements in $G_1 \text{ x } G_2$ and k_1, k_2 are elements in $K_1 \text{ x } K_2$.

Then, we need to show:

$$f[(g_1,g_2)(g_1',g_2')] = f[(g_1,g_2)]f[(g_1',g_2')].$$

$$\text{So, } f[(g_1,g_2)(g_1',g_2')] = [(g_1k_1,\ g_2k_2)][(g_1'k_1,\ g_2'k_2)]$$

$$= [(g_1g_1'k_1,\ g_2g_2'k_2)]$$

$$= g_1g_1'k_1 \times g_2g_2'k_2$$

$= G_1 / K_1 \times G_2 / K_2.$

3.3.1 Homomorphism and Normal Subgroups

It often happens that some subset of a group will also form a group under the same operation. Such a group is called a subgroup. If (G, •) is a group and H is a nonempty subset of G, then (H,•) is called a subgroup of (G,•) if the following conditions exists:

- a• b ∈ H for all a, b ∈ H. (closure)
- a− 1 ∈ H for all a ∈ H. (existence of inverses).

Homomorphism

If (G,•) and (H, *) are two groups, the function f :G → H is called a group homomorphism if f (a• b) = f (a) * f (b) for all a, b ∈ G.

We often use the notation f: (G,•) → (H, *) for such a homorphism. Many authors use morphism instead of homomorphism. A group isomorphism is a bijective group homomorphism. If there is an isomorphism between the groups (G,•) and (H, *), we say that (G,•) and (H, *) are isomorphic and write (G, •) ≅ (H, *).

Semi Group Homomorphism

If {S, *} and {T, Δ} are any two semi groups, then the mapping f: S → T such that, for any two elements a, b ∈ S.

f (a * b) = f(a) Δ f(b) is called a semi group homomorphism.

The composition of semi group homomorphism is also a semi group homomorphism.

If {S, *}, {T, Δ} and {V, ⊕} are semi groups and if f: S → T and g: T → V are homomorphism, then To prove that g • f: S → V is also a homomorphism.

Let a, b ∈ S. Then:

$$(g \cdot f)\ (a * b) = g\ \{f(a * b)\}$$

$$= g\ \{f\ (a)\ \oplus\ f\ (b)\}$$

$$= g\,f\ (a)\ \oplus\ gf\ (b)$$

$$= (g \cdot f)\ (a)\ \oplus\ (g \cdot f)\ (b).$$

i.e., g • f: S → V is also a homomorphism.

i.e., The composition of semi group homomorphism is also a semi group homomorphism.

Problems

1. Let us calculate a sub-group of order two of the group (Z_8, +8).

Solution:

Given:

$$\left(Z_8,\ +8\right)$$

By Cayley's table:

{[0], [4]} is the only sub-group of order 2.

$+_8$	[0]	[4]
[0]	[0]	[4]
[4]	[4]	[0]

2. Let us calculate all the subgroups of (z_{12}, +12).

Solution:

Given:

$$\left(z_{12},+12\right)$$

Cayley's table of (Z_{12}, + 12).

$+_{12}$	0	1	2	3	4	5	6	7	8	9	10	11
0	0	1	2	3	4	5	6	7	8	9	10	11
1	1	2	3	4	5	6	7	8	9	10	11	0
2	2	3	4	5	6	7	8	9	10	11	0	1
3	3	4	5	6	7	8	9	10	11	0	1	2
4	4	5	6	7	8	9	10	11	0	1	2	3
5	5	6	7	8	9	10	11	0	1	2	3	4
6	6	7	8	9	10	11	0	1	2	3	4	5
7	7	8	9	10	11	0	1	2	3	4	5	6
8	8	9	10	11	0	1	2	3	4	5	6	7
9	9	10	11	0	1	2	3	4	5	6	7	8
10	10	11	0	1	2	3	4	5	6	7	8	9
11	11	0	1	2	3	4	5	6	7	8	9	10

From the Cayley's table,

The identity element is '0'

Inverse of the elts 0 is 0, 1 is 11, 2is 10.

3 is 9, 4 is 8, 5 is 7

6 is 6, 7 is 5, 8 is 4

9 is 3, 10 is 2 & 11 is 1.

From the Cayley's table.

The subgroup of (Z_{12}, + 12) are as follows:

$(1)\left[\{[0],[6]\}+_{12}\right]$

$(2)\left[\{[0],[3],[6],[9]\}+_{12}\right]$

$(3)\left[\{[0],[2],[4],[6],[8],[10]\}+_{12}\right]$

$(4)\left[\{[0]\},_{12}\right]$

$(5)[Z_{12}+_{12}]$

The group itself is a subgroup and the identity element together with binary operation is also a subgroup, (trivial subgroup).

3. Let G be a group and a ∈ G. Let us show that the subset $\{axa^{-1} : x \in G\}$ of G is the subgroup of G.

Solution:

Given:

$a \in G$ and $H = \{a\,x\,a^{-1} \mid x \in G\}$

Now, $a = a$

$aa^{-1} \in H$

∴ H is not empty.

Let b a xa and $c = a\,ya^{-1} \in H$.

Then, $bc^{-1} = (ax\,a^{-1})(a\,ya^{-1})^{-1}$

$$= a\;xa^{-1}a\;y^{-1}a^{-1}$$

$$= axe\;y^{-1}a^{-1}$$

$$= a(xy^{-1})\;a^{-1} \in H.$$

So, $bc^{-1} \in H$ for all $b, c \in H$.

Hence, H is a subgroup of G.

4. If G_1 and G_2 are groups and f: $G_1 \rightarrow G_2$ is a homomorphism, let us prove that the kernel of f viz, ker f is a normal sub-group of G_1.

Solution:

Given:

Let f: $G_1 \rightarrow G_2$ be a group homomorphism and ker f = K = x ∈ G_1 I f(x) = e be the kernel f 'e' being the identify element of G_2.

As f (e) = e we have e ∈ k and k ≠ φ

Let x, y ∈ k Then f (xy^{-1}) = f (x) f (y^{-1})

$$= f(x)\;(f(y))^{-1}$$

$$= e' \text{ in } G_2$$

And hence xy^{-1} ∈ k.

Thus, xy^{-1} ∈ k for all x, y ∈ k and hence k is a subgroup of G_1.

To prove that k is normal:

Let a ∈ G_1 and x ∈ k then f(x) = e′ and

$$F\;(a\,x\,a^{-1}) = f(a) \bullet f(x) \bullet f(a^{-1}) \therefore \text{f is homomorphism}$$

$$= f(a)\;e' \bullet (f(a))^{-1}$$

$$= f(a)\;(f(a))^{-1} \text{ as } f(x) = e' \text{ and } f(a^{-1}) = (f(\,a))^{-1}$$

$$= e'$$

Thus, $axa^{-1} \in k$ for all $a \in G_1$ and $x \in k$.

Hence, k is a normal subgroup of G_1.

5. Let us prove that the Kernel of a homomorphism of a group (G, *) into an another group (H, Δ) is a subgroup of G.

Solution:

Given:

Let f: (G, *) → (H,Δ) is a group homomorphous,

WKT f (e) = e, where e and e are the identifies of G and H.

∴ e ∈ ker (f).

ker (f) is a non-empty subset of (G, *).

Let a, b ∈ ker (f).

F (a) = e and f (b) = e by definition.

Now, $f\left(a * b^{-1}\right) = f(a) \Delta f\left(b^{-1}\right)$

$= f\,(a)\left(\Delta f\,(b)\right)^{-1}$

$= e' \Delta \left(e'\right)^{-1}$

$f\left(a * b^{-1}\right) = e' \Delta e$

$= e'$

$\therefore a * b^{-1} \in \text{ker}\ (f)$

Thus, when a, b, ∈ ker (f); a * b^{-1} ∈ ker (f).

Hence, ker (f) is a subgroup of (G, *).

6. If S = N x N, the set of ordered pairs of positive integer with the operation * defined by (a, b) * (c, d) = (ad + b c, b d) and if f: (S, *) → (Q, +) is defined by f (a, b) = a/b, then Let us show that f is a semi-group homomorphism.

Solution:

Given:

$$\left\{(a,b) * (c,\ d)\right\} * (e,f) = (ad + bc, bd) * (e,\ f)$$

$=\{(ad+bc)f+bde,\ bdf\}$

$=(adf+bcf+bde, bdf)$

Also $(a,b)*\{(c,d)*(e,\ f)\}=(a,b)*(cf+de,df)$

$=\{adf+b(cf\ +\ de),\ bdf\}$

$=(adf+bcf+bde,bdf).$

Hence, (S, *) is associative and hence a semi group.

Now $f\ ((a,\ b)*(c,\ d))=f(ad+b\,c,\ b\,d)$

$=\frac{ad+bc}{bd}\left[\because f(a,b)=\frac{a}{b}\right]$

$=\frac{a}{b}+\frac{c}{d}$

$=f(a,\ b)\ +\ f(c,\ d)$

$\therefore f:(S,*)\rightarrow(Q,+)$ is a semi group homomorphism.

7. Let us show that the intersection of two normal subgroups of a group G is also a normal subgroup of G.

Solution:

Given:

Let H and K are normal subgroups of G.

To prove that $H \cap K$ is also a normal subgroup of G.

Claim: $H \cap K$ is a subgroup of G.

Let a, b $\in H \cap K \Rightarrow$ a, b $\in$ H and a, b $\in$ K

$\Rightarrow ab^{-1} \in H$ and $ab^{-1} \in K$ ($\because$ H & K are subgroup)

$\Rightarrow ab^{-1} \in H \cap K$

Hence, $H \cap K$ is a subgroup of G.

To prove that $H \cap K$ is normal subgroup of G.

Let $h \in H n K$ and $g \in G$.

$\Rightarrow h \in H$ and $g \in G$ and $h \in K$ and $g \in G$.

$\Rightarrow ghg^{-1} \in H$ and $ghg^{-1} \in K$ ($\because$ H & K are normal)

$\Rightarrow ghg^{-1} \in H \cap K$

Hence, $H \cap K$ is normal subgroup of G.

Hence, Intersection of two normal sub groups is normal.

8. If H is a subgroup of G such that $x^2 \in H$ for every $x \in G$. Let us prove that H is a normal subgroup of G.

Solution:

Given:

$x_2 \in H$ for every $x \in G$

For any $a \in G$ and $h \in H$, we have $a * h \in G$, by closure property.

$\therefore (a * h)^2 \in H$, by the given condition ...(1)

Also since $a^{-1} \in G$, $(a^{-1})^2 = a^{-2} \in H$, by the given condition. Since H is a subgroup and h^{-1}, $a^{-2} \in H$, we have:

$h^{-1} * a^{-2} \in H$ ...(2)

From (1) and (2) we have:

$$(a * h)^2 * h - 1 * a - 2 \in H$$

$$\text{i.e.,} a * h * a * h * h^{-1} * a^{-2} \in H$$

$$\text{i.e.,} a * h * a * e * a^{-2} \in H, \text{ where e is the identity}$$

$$\text{i.e..} \ a * h * a^{-1} \in H$$

$$\text{or } a^{-1} * h * a \in H \left(\text{by replacing a by } a^{-1}\right)$$

$\therefore$ H is a normal subgroup.

3.3.2 Rings

In mathematics, a ring is an algebraic structure consisting of a set together with two binary operations usually called addition and multiplication, where the set is an abelian group under addition (called the additive group of the ring) and a monoid under multiplication such that multiplication distributes over addition.

In other words, the ring axioms require that addition is commutative, addition and multiplication are associative, multiplication distributes over addition, each element in the set has an additive inverse, and there exists an additive identity. One of the most common examples of a ring is the set of integers endowed with its natural operations of addition and multiplication.

A ring is a set of R on which there are defined two binary operations + and x, satisfying the following axioms.

R1: R with the operation + is a commutative group.

R2: The operation x has the closure, associativity and identity properties.

R3: (The distributive laws). For all a, b and c in R:

$$a \times (b + c) = (a \times b) + (a \times c)$$

$$(a + b) \times c = (a \times c) + (b \times c)$$

Commutative Ring

If (R, X) is commutative, then the ring (R, +, X) is called a commutative ring.

Commutative Ring with Identity

- $(Z_4, +_4, X_4)$ is a commutative ring with identity.
- (Z, ⊕, •) is a commutative ring with identity.

Where the operations ⊕ and • are defined, for any a, b ∈ Z and a ⊕ b = a + b + 1 and a • b = a + b + ab.

3.3.3 Integral Domains and Fields

These are two special kinds of rings.

Zero-divisors

If a, b are two ring elements with a, b ≠ 0, but a b = 0, then a and b are called zero-divisors.

Example:

In the ring Z_6, we have 2.3 = 0 and so, 2 and 3 are zero-divisors.

More generally, if n is not prime and then Z_n contains zero-divisors.

Integral Domain

An integral domain is a commutative ring with an identity ($1 \neq 0$) with no zero-divisors. That is $a b = 0 \Rightarrow a = 0$ or $b = 0$.

Examples:

- The ring Z is an integral domain.
- The polynomial rings Z[x] and R[x] are integral domains.
- The ring $\{a + b\sqrt{2} \mid a, b \in Z\}$ is an integral domain.
- If p is prime then ring Z_p is an integral domain.

Field

A field is a commutative ring with identity ($1 \neq 0$) in which every non-zero element has a multiplicative inverse.

Examples:

- The rings Q, R, C are fields.

Theorem:

Every finite integral domain is a field.

Proof:

The only thing we need to show is that a typical element $a \neq 0$ has a multiplicative inverse. Consider $a, a^2, a^3, \ldots$ Since there are only finitely many elements, we must have $a^m = a^n$ for some $m < n$(say). Then $0 = a^m - a^n = a^m (1 - a^{n-m})$.

Since there are no zero-divisors, we must have $a^m \neq 0$ and hence $1 - a^{n-m} = 0$ and so $1 = a(a^{n-m-1})$ and we have found a multiplicative inverse for a.

3.4 Boolean Algebras: Lattices and Algebraic Systems

A Boolean function is a special kind of mathematical function $f: X^n \rightarrow X$ of degree n, where $X = \{0, 1\}$ is a Boolean domain and n is a non-negative integer. It describes the way how to derive Boolean output from Boolean inputs.

Example: Let, F (A, B) = A'B'. This is a function of degree 2 from the set of ordered pairs of Boolean variables to the set {0, 1} where F (0, 0) = 1, F (0, 1) = 0, F (1, 0) = 0 and F (1, 1) = 0.

Algebraic System

An algebraic system is defined as a set along with operations on elements of the set. An algebraic structure is defined as set operations on elements of the set and relations between elements of the set which leads to a structure on the elements of a set.

An algebraic system is a system consisting of a nonempty set A and one or more n-ary operations on the set A. It is denoted by (A, f_1, f_2,...).An algebraic structure is an algebraic system, (A,f_1,f_2,...,R_1,R_2,...), wherein addition to operations fi, the relations Ri are defined on A. This leads to a structure on the elements of A.

Boolean Lattice

A Boolean lattice B is a distributive lattice in which for each element x ∈ B in B there exists a complement x ∈ B such that:

$$x \wedge x' = 0$$

$$x \vee x' = 1$$

$$(x')' = x$$

$$(x \wedge y)' = x' \vee y'$$

$$(x \vee y)' = x' \wedge y'$$

Properties of Lattices

If {L, ≤} is a lattice, then f or any a, b, c ∈ L:

- $a \vee a = a$; $a \wedge a = a$ (Idempotency).
- $a \vee b = b \vee a$; $a \wedge b = b \wedge a$ (Commutative).

Absorption laws of Lattice:

If a 仟 (a + b) = a,

then, a + (a 仟 b) = a for all a,b ∈ L

Proof:

* a • (a + b) = (a + o) (a + b); by identity laws

= a + (o • b); Distributive law

= a + 0; by dominance law

a • (a + b) = a; identity law.

* Now a + (a • b) = (a • 1) + (a • b); identity law

= a • (1 + b); distributive law

= a • 1; by dominance law

a + (a • b) = a; identity law.

Hence proved.

Lattices as Algebraic Systems - Sub Lattices

A non-empty subset M of a lattice (L, V, ∧) is called a sub lattice of L, if M is closed under both the operations V and ∧.

Problems

1. Let us prove that in a Boolean algebra, $x + x \cdot y = x$.

Solution:

$x + x \cdot y = x \cdot 1 + x \cdot y$, by identity law

$= x \cdot (1 + y)$, by distributive law

$= x \cdot 1$, by dominance law

$= x$, by identity law.

2. Give an example of two-element Boolean algebra.

Solution:

Let (B, *, °, 0, 1) be an example of two element Boolean algebra where B = {0, 1}.

3. In any Boolean algebra, let us show that $(a + b)(a + c) = (ac + ab) = ac + ab + bc$.

Solution:

Now,

$(a + b)(a' + c) = aa' + ac + a'b + bc$

$= 0 + ac + a'b + bc \because aa' = 0$

$= ac + a'b + bc$ ...(1)

Now,

$ac + a'b + bc = ac + a'b + bc(a + a')$

$= ac + a'b + abc + a'bc$

$= (ac + abc) + (a'b + a'bc)$

$= ac(1 + b) + a'b(1 + c)$

$= ac + a'b$ ...(2)

From (1) & (2),

$(a + b)(a' + c) = ac + a'b = ac + a'b + bc.$

4. Let us show that the complement of every element in a Boolean algebra is unique.

Solution:

Let a ∈ L has two complement b and c ∈ L.

By definition, a * b = 0; a ⊕ b = 1 and

$a * c = 0; a \oplus c = 1$ ...(1)

Now b = b * 1; identify law

$= b * (a \oplus c)$; by (1)

$= (b * a) \oplus (b * c)$; By distributive law

$= (a * b) \oplus (b * c)$; Commutative law

$= 0 \oplus (b * c)$

$= (a * c) \oplus (b * c)$; by (1)

$= (a \oplus b) * c$; by distributive law

$= 1 * c$

$b = c.$

Hence, every element of L has a unique complement.

5. Let us show that in a Boolean algebra a b′ + a′ b = 0 if and only if a = b.

Solution:

Let $ab' + a'b = 0.$

Then $a + ab' + a'b = a$

$a + ab = a$; by absorption law

$(a + a') \cdot (a + b) = a$

$1 \cdot (a + b) = a$

$a + b = a$ ...(1)

Similarly,

$ab' + a'b + b = 0$

$ab' + b = b$

$(a + b) \cdot (b + b) = b$

$(a + b) \cdot 1 = b$

$a + b = b$ ...(2)

From (1) & (2),

$a = b$

Then, $ab' + a'b = aa' + a'a$

$= 0 + 0$

$= 0$

6. Let us prove that D_{110}, the set of all positive divisors of a positive integer 110 is a Boolean algebra and find all its sub algebras.

Solution:

Let $D_{110} = \{1, 2, 5\ 10, 11, 22, 55, 110\}$.

Hasse diagram of $(D_{110}, /)$ is

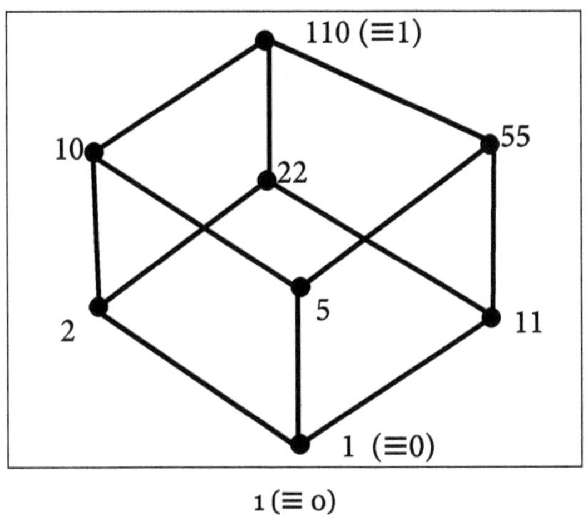

1 (≡ 0)

From the diagram, every pair of element has lub and glb.

Here least element is 1 (= 0)

Greatest element is 110 (= 1)

Also the binary operation $\vee$, $\wedge$ satisfy distributive property.

Hence, $(D_{110}, /)$ is a distributive lattice i.e., the zero element of the lattice is 1. And the unit element of the lattice is 110.

Also $1 \vee 110 = \text{LCM}(1, 110) = 110 \equiv 1$

$1 \wedge 110 = \text{GCD}(1, 110) = 1 \equiv 0$

$\therefore$ Complement of 1 is 110.

Similarly,

LCM (2, 55) = 110; GCD (2, 55) = 1

$\therefore$ Complement of 2 is 55.

LCM (10, 11) = 110; GCD (10, 11) = 1

$\therefore$ Complement of 10 is 11.

LCM (5, 22) = 110; GCD (5, 22) = 1

$\therefore$ Complement of 5 is 22.

Hence, complement of each and every element exist.

$\therefore (D_{110}, /)$ is a Boolean algebra.

Sub algebras are:

$$\{1,110\},\ \{1, 2, 55, 110\},\ \{1, 5, 22, 110\},\ \{1, 10, 11, 110\}.$$

7. In a Boolean algebra B, let us prove that $(a \wedge b)' = a' \vee b'$ and $(a \vee b)' = a' \wedge b'$ for all $a, b \in B$.

Solution:

Since the lattice is complemented.

The complements of a and b exist.

Let Complement of a = a′

$b = b'$.

Now, $(a \wedge b) \vee (a' \vee b') = (a \vee (a' \vee b')) \wedge (b \vee (a' \vee b'))$; by distributivity

$= ((a \vee a') \vee b') \wedge ((b \vee b') \vee a')$; by associating and commutability

$= (1 \vee b') \wedge (1 \vee a')$

$= | \wedge |$

$= 1$ ...(1)

$(a \wedge b) \wedge (a' \wedge b') = ((a \wedge b) \wedge a') \vee ((a \wedge b) \wedge b')$

$= ((a \wedge a') \wedge b) \vee (a \wedge (b \wedge b'))$

$= (0 \wedge b) \vee (a \wedge 0)$

$= 0 \vee 0$

$= 0$ ...(2)

From (1) & (2), we get, a’ ∨ b’ is the complement of $(a \wedge b)$

i.e., $(a \wedge b)' = a' \vee b'$ ...(3)

By the principle of duality, it follows from (3)

$(a \vee b)' = a \wedge b'.$

8. If KB is a Boolean algebra, then for a ∈ B a + 1 = 1, a • 0 = 0.

Solution:

Dominance Laws:

a + 1 = 1 and a • 0 = 0, for all a ∈ B

a + 1 = (a+ 1) • 1, by Identity Laws

= (a + 1) (a + a), by Complement Laws

= a + (1 • a), by Distributive Laws

= a + (a • 1), by Commutative Laws

= a + a , by Identity Laws

= 1, by Complement Laws.

Now, a • 0 = (a • 0) + 0, by Identity Laws

= (a • 0) + (a • a´), by Complement Laws

= a • (0 + a´), by Distributive laws

= a • (a´ + 0), by Commutative Laws

= a • a, by Identity Laws

= 0, by Complement Laws.

Hence proved.

Lattices

9. Let us show that in a lattice, $a \le b \Rightarrow a * b = a$.

Solution:

Given:

$a \le b$

W.K.T a ≤ a

Clearly, $a * b \le a$ ...(1)

Since a * b is a lower bound of 'a'.

Now, a is a lower bound of a & b. ...(2)

Since a * b is the greatest lower bound of a and b. ...(3)

From (2) & (3):

$$a \le a * b \quad ...(4)$$

From (1) & (4), we get:

$$a * b = a$$

Hence in a lattice, $a \le b \Rightarrow a * b = a$.

10. Let us prove that in a lattice (L, ≤) for any a, b, c ∈ L, a ⊕ (b * c) ≤ (a ⊕ b) * (a ⊕ c).

Solution:

Given:

A partially ordered set {L, ≤} in which every pair of elements has a least upper bound and a greatest lower bound is called a Lattice.

Since a ⊕ b is least upper bound (L ∪ B) of (a, b)

$$\therefore a \le (a \oplus b) \quad ...(1)$$

Also, $(b * c) \le b \le (a \oplus b)$ ...(2)

Since (b * c) is the greatest lower bound (GLB) of b & c From (1) and (2) we get:

(a ⊕ b) is a upper bound of {a, b * c} ...(3)

$$\therefore a \oplus (b * c) \le (a \oplus b) \quad ...(4)$$

Similarly,

$$a \le (a \oplus C) \quad ...(5)$$

Also, $(b * c) \le C \le (a \oplus c)$ ...(6)

Since b * c is the GLB of b & c.

From (5) & (6):

(a ⊕ c) is a upper bound of {a, b * c}

$$a \oplus (b * c) \le (a \oplus c) \quad ...(7)$$

From (3) & (7), we get:

$$a \oplus (b * c) \leq (a \oplus b) * (a \oplus c).$$

11. Let us show that the direct product of any two distributive lattices is a distributive lattice.

Solution:

Given:

Let (L, +, *) and (S, ∧, ∨) be two distributive lattice with respect to relation ≤ and ≤ respectively.

Then + and '*' in L × S is defined by:

$$(a,\ b) + (c,\ d) = (a * c,\ b \vee d)$$

$$(a,\ b) * (c,\ d) = (a * c,\ b \wedge d).$$

A relation α on L × S is defined by:

$$(x,\ y)\ \alpha\ (z,\ u) \Leftrightarrow x \leq z;\ y \leq u.$$

To prove that L × S is a distributive lattice, we have to prove that:

- L × S is a poset.
- L × S is a lattice.
- L × S is a distributive lattice.

Case (i)

Let $(a,\ b) \in L \times S$

$$\Rightarrow\ a \in L,\ b \in S$$

$$a \leq a,\ b \leq b.$$

$$\therefore (a,b)\alpha(a,b)$$

∴ α is reflexive in L × S.

Let $(a,b),(c,d) \in L \times S$ such that:

$$(a,b)\alpha(c,d) \text{and} (c,d)\alpha(a,b) \Rightarrow a \leq c, b \leq d \text{ and } c \leq a, d \leq b.$$

$\therefore$ Now $a \leq c,\ c \leq a \Rightarrow a = c.$

$b \leq d,\ d \leq b \Rightarrow b = d.$

$\therefore$ (a, b) = (c, d). Hence, α is anti-symmetric.

Let $(a_1, b_1), (a_2, b_2), (a_3, b_3) \in L \times S$ such that:

$$(a_1, b_1) \propto (a_2, b_2), (a_2, b_2) \propto (a_3, b_3)$$

$\therefore\ a_1 \leq a_2;\ b_1 \leq b_2$

$a_2 \leq a_3;\ b_2 \leq b_3$

Now, $a_1 \leq a_2,\ a_2 \leq a_3 \Rightarrow a_1 \leq a_3$

$b_1 \leq b_2,\ b_2 \leq b_3 \Rightarrow b_1 \leq b_3$

$\therefore\ (a_1, a_3)\ \alpha\ (b_1, b_3)$

$\therefore$ α is transitive.

Hence, the relation α is a posset on L × S.

Case (ii)

To prove that L × S is lattice.

Let $(a, b), (c, d) \in L \times S$

$\Rightarrow\ a, c \in L;\ b, d \in S.$

Since L is a lattice, a and c have lub $u_1 \in L$ also a and c have glbl, ∈ L, b, d∈ S.

$\therefore$ b and d have lub $u_2 \in S$, b and d have glb $l_2 \in S$.

Then (a, b) and (c, d) have lub (u_1, u_2) and (a, b) * (c, d) have glb $(l_1 l_2)$ by definition.

$\therefore$ L × S is a lattice.

Case (iii)

To prove that L × S is a distributive lattice.

Let $x = (a_1, b_1); y = (a_2, b_2); z = (a_3, b_3) \in L \times S$

$\Rightarrow a_1 a_2, a_3 \in L$ and $b_1 b_2, b_3 \in S$.

Now,

$$x * (y + z) = (a_1, b_1) * [(a_2, b_2) + (a_3, b_3)]$$

$$= (a_1, b_1) * [a_2 \vee a_3, b_2 \vee b_3]$$

$$= [(a_1 * (a_2 \vee a_3)], b_1 \wedge (b_2 \vee b_3]$$

$$= [(a_1 * a_2) \vee (a_1 * a_3), (b_1 \wedge b_2) \vee (b_1 \wedge b_3)]$$

$$= [(a_1 * a_2), (b_1 \wedge b_2)] + [(a_1 * a_3), (b_1 \wedge b_3)]$$

$$= (a_1, b_1) * (a_2, b_2) + (a_1 b_1) * (a_3 b_3)$$

$$= x * y + x * z.$$

i.e., '*' is distributive over '+'.

Now,

$$x + (y * z) = (a_1, b_1) + ((a_2, b_2) * (a_3, b_3))$$

$$= (a1, b_1) + (a_2 * a_3 * b_2 \wedge b_3)$$

$$= [a_1 \oplus (a_2 * a_3), b, \vee (b_2 \wedge b_3)]$$

$$= [(a_1 \oplus a_2) * (a_1 \oplus a_3), (b_1 \wedge b_2) \wedge (b_1 \vee b_3)]$$

$$= [(a_1 \oplus a_2), b_1 \vee b_2] * [(a_1 \oplus i_3), (b_1 \vee b_3)]$$

$$= [(a_1, b_1) + (a_2, b_2)] * [(a_1, b_1) + (a_3, b_3)]$$

$$= (x + y) * (x + z).$$

+ is distributive over '*'.

Hence, (L × S, *, +) is distributive lattice with relation α.

12. In a Lattice (L, ≤), let us prove that $X \vee (Y \wedge Z) \leq (X \vee Y) \wedge (X \vee Z)$.

Solution:

Given:

(L, ≤)

Let x, y, z ∈ L.

$$x \leq (x \vee y) \,\&\, x \leq (x \vee z) \Rightarrow x \leq (x \vee Y) \wedge (x \vee z) \quad ...(1)$$

$$(Y \wedge Z) \leq Y \,\&\, Y \leq (X \vee Y) \text{ i.e., } (Y \wedge Z) \leq (X \vee Y) \quad ...(2)$$

$$(Y \wedge Z) \leq Z \,\&\, Z \leq (X \vee Z) \text{ i.e., } (Y \wedge Z) \leq (X \vee Z) \quad ...(3)$$

From (2) & (3):

$$(Y \wedge Z) \leq (X \vee Y) \wedge (X \vee Z) \quad ...(4)$$

From (1) & (4):

$$X \vee (Y \wedge Z) \leq (X \vee Y) \wedge (X \vee Z)$$

Hence proved.

3. Let us prove the distributive inequalities of lattices.

Solution:

Given:

If { L, ≤} is a lattice, then for any a, b , c ∈ L:

$$*a \wedge (b \vee c) \geq (a \wedge b) \vee (a \wedge c)$$

$$*a \vee (b \wedge c) \leq (a \vee b) \wedge (a \vee c)$$

Proof:

Since $a \wedge b$ is the greatest lower bound of (a, b)

$$\therefore\ a \wedge b \leq a \quad ...(1)$$

Also, $a \wedge b \leq b \leq b \vee c$ ...(2)

Since $b \vee c$ is the least upper bound of (b, c).

From (1) & (2), we have:

(a ∧ b) is a lower bound of (a, b ∨ c)

$$(a \wedge b) \leq a \wedge (b \vee c) \qquad ...(3)$$

Similarly, $(a \wedge c) \leq a$

And, $(a \wedge) \leq c \leq b \vee c$

$$\therefore\ (a v c) \leq a \wedge (b \vee c) \qquad ...(4)$$

From (3) & (4), we get:

$$(a \wedge b) \vee (a \wedge c) \leq a \wedge (b \vee c)$$

i.e., $a \wedge (b \vee c) \geq (a \wedge b) \vee (a \wedge c)$

By the principle of duality:

$$a \vee (b \wedge c) \geq (a \vee b) \wedge (a \vee c).$$

Hence proved.

3.4.1 Principle of Duality

The Duality Theorem starts with a Boolean relation. We can derive another Boolean relation by:

- Changing OR (operation) i.e., + (Plus) sign to an AND operation.
- Complement any 0 or 1 appearing in the expression.

Duality Law

Two formulas, A and A*, are said to be duals of each other if any one of them can be obtained from the other by replacing Λ by V and V by Λ. The connectives Λ and V are called duals of each other. If the formula A contains the special variables T and F, then A*, the dual of A, is obtained by replacing T by F and F by T in addition to the previously mentioned interchanges.

Theorem: Let A and A* be dual formulas. If $P_1, P_2, ...$ and P_n are all the atomic variables that occur in A and A*, then:

$$\neg A(P_1, P_2, ..., P_n) \Leftrightarrow A^*(\neg P_1, \neg P_2, ... \neg P_n)$$

And, $A(\neg P_1, \neg P_2, ..., \neg P_n) \Leftrightarrow A^*(P_1, P_2, ... P_n).$

Proof:

For n atomic variables we have:

$$(P_1, \vee P_2 \vee, \ldots, \vee P_n) \Leftrightarrow \neg(\neg P_1 \wedge \neg P_2 \wedge \ldots \neg P_n)$$

And $(P_1 \wedge P_2 \wedge \ldots \wedge P_n) \Leftrightarrow \neg(\neg P_1 \vee \neg P_2 \vee \ldots \vee \neg P_n)$

Therefore we have, $A(P_1, P_2, \ldots, P_n) \Leftrightarrow \neg A^*(\neg P_1, \neg P_2, \ldots, \neg P_n)$.

Negation of $A(P_1, P_2, \ldots, P_n) \Leftrightarrow \neg A^*(\neg P_1, \neg P_2, \ldots, \neg P_n)$ gives $\neg A(P_1, P_2, \ldots, P_n) \Leftrightarrow A^*$ $(\neg P_1, \neg P_2, \ldots, \neg P_n)$.

If we replace Pi by ¬Pi for i = 1….n, then we have:

$$A(\neg P_1, \neg P_2, \ldots, \neg P_n) \Leftrightarrow A^*(P_1, P_2, \ldots, P_n).$$

Negation of $\neg A(\neg P_1, \neg P_2, \ldots, \neg P_n) \Leftrightarrow A^*(P_1, P_2, \ldots, P_n)$ gives $A(\neg P_1, \neg P_2, \ldots, \neg P_n) \Leftrightarrow \neg A^*$ $(P_1, P_2, \ldots, P_n)$.

3.4.2 Distributive and Complemented Lattices

Distributive Lattice

A lattice A is distributive if the following distributive law exists for any x, y, and z in $A: x \wedge (y \vee z) = (x \wedge y) \vee (x \wedge z)$; $x \vee (y \wedge z) = (x \vee y) \wedge (x \vee z)$.

We already know that meet and join are commutative, so if a lattice satisfies the above distributive laws, it automatically satisfies the other distributive laws as well.

$$(x \wedge y) \vee z = (x \vee z) \wedge (y \vee z) \text{ and } (x \vee y) \wedge z = (x \wedge z) \vee (y \wedge z)$$

Complemented Lattices

A bounded lattice is complemented if every element has a complement. Complemented lattices are, by definition, lattices with complements for all their elements; but the existence of complements still does not guarantee uniqueness. Nevertheless, if the lattice is distributive as well, this leeway disappears, as the next proposition demonstrates. The corollary that follows is immediate.

Problem

Let us show that the following simple but significant lattices are not distributive.

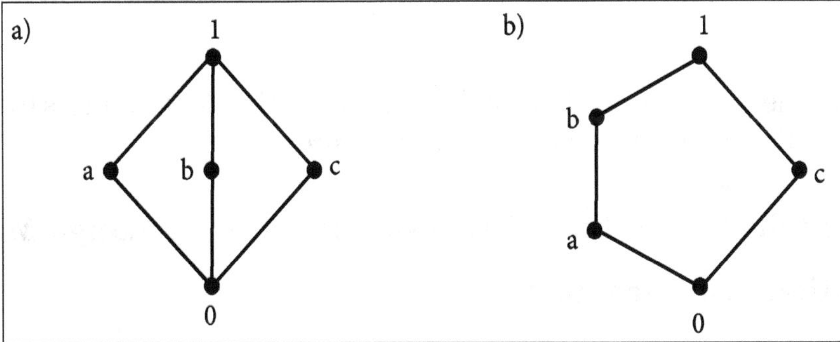

Solution:

Given:

- To see that the diamond lattice is not distributive, use the middle elements of the lattice: $a \wedge (b \vee c) = a \wedge 1 = a$, but $(a \wedge b) \vee (a \wedge c) = 0 \vee 0 = 0$, and a 6= 0. Similarly, the other distributive law fails for these three elements.
- The pentagon lattice is also not distributive.

3.5 Boolean Functions: Boolean Expressions

A Boolean expression in the n variables $x_1, x_2, \ldots, x_n$, is any expression $E(x_1, x_2, \ldots, x_n)$ formed from these variables and the constants 0 and 1 via the operations of Boolean Algebra. More explicitly, Boolean expressions in $x_1, x_2, \ldots, x_n$ are defined recursively as follows:

1) Constants: 0 and 1 are Boolean expressions in $x_1, x_2, \ldots, x_n$.

2) Variables: $x_1, x_2, \ldots, x_n$ are Boolean expressions in $x_1, x_2, \ldots, x_n$.

3) Sums, Products and Complements: If E and F are Boolean expressions in $x_1, x_2, \ldots,$ x_n, then so are $E+F$, $E \cdot F$, and $\bar{E}$.

Thus, E is a Boolean expression in $x_1, x_2, \ldots, x_n$ if it has been generated in a finite number of steps from the constants and variables using sums, products or complements.

Problems

Let us determine the simple Boolean formula for the function $f : B^2 \rightarrow B$ defined by $f(x, y) = x + xy$.

Solution:

Given:

$$f : B^2 \rightarrow B$$

$$f(x, y) = x + xy$$

According to one of the Absorption Laws, x + x y = x. A simpler formula is thus given by f(x, y) = x; f is the projection map onto the first coordinate.

3.5.1 Simplification of Logic Expressions using Karnaugh Map

Construction of a Karnaugh Map

An n-variable Karnaugh map has 2n squares and each possible input is allocated a square. In the case of a min term Karnaugh map, '1' is placed in all those squares for which the output is '1' and '0' is placed in all those squares for which the output is '0'. 0s are omitted for simplicity. An 'X' is placed in squares corresponding to don't care'conditions.

In the case of a max term Karnaugh map, a '1' is placed in all those squares for which the result is '0' and a '0' is placed for input entries corresponding to a '1' result. Then, 0s are omitted for simplicity and an 'X' is placed in squares corresponding to don't care' conditions. The choice of terms identifying different rows and columns of a Karnaugh map is not unique for a given number of variables.

The only condition to be satisfied is that the designation of adjacent rows and adjacent columns should be the same except for one of the literals being complemented. Also, the extreme rows and extreme columns are considered adjacent. Some of the possible designation styles for 2, 3 and 4 variable min term, Karnaugh maps are shown in the figure below.

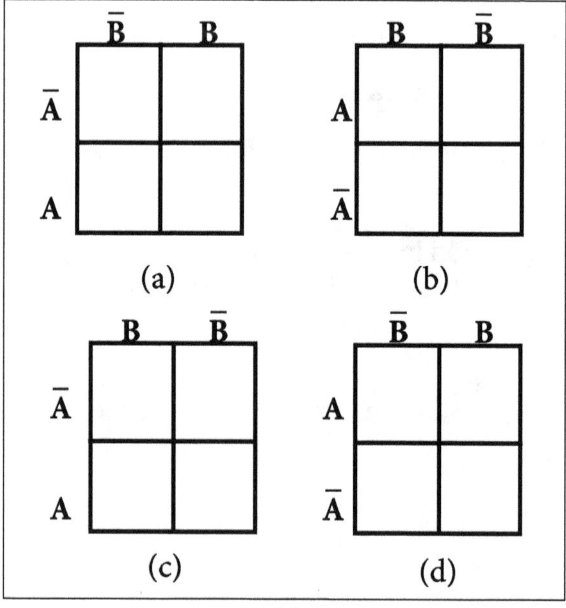

Two variable K Map.

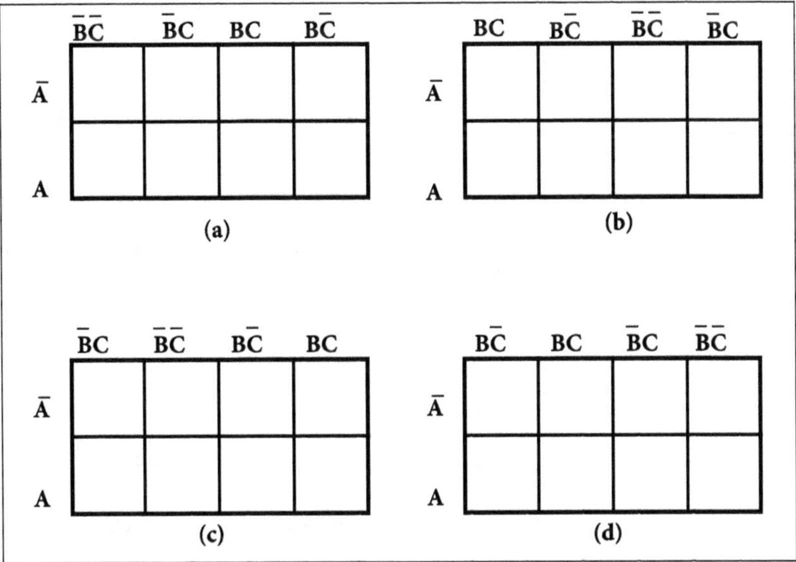

Three variable K Map.

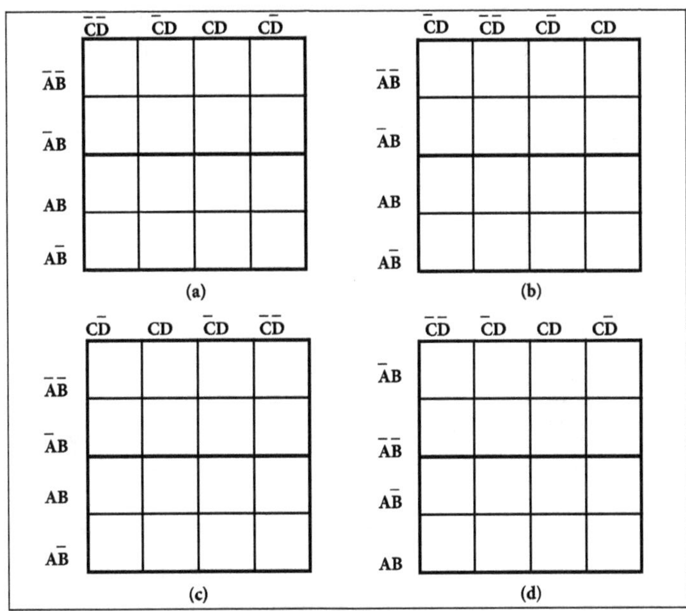

Four variable K Map.

The style of row identification need not be the same as that of column identification as long as it meets the basic requirement with respect to adjacent terms. It is, however, accepted practice to adopt a uniform style of row and column identification. Also, the style shown in the figure below is more commonly used. Having drawn the Karnaugh map, the next step is to form groups of 1s as per the subsequent guidelines:

- Each square containing a '1' must be considered at least once, although it can be considered as often as desired.

- The goal should be to account for all the marked squares in the minimum number of groups.
- The number of squares in a group should always be a power of 2, i.e. groups can have 1, 2, 4, 8, 16, squares.

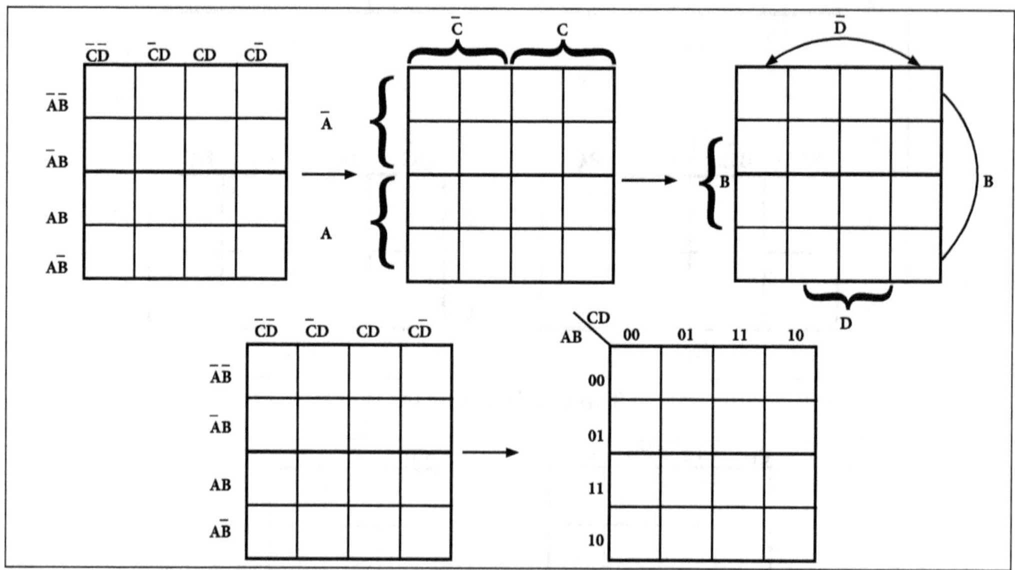

Different Styles of row and column identification.

4. Each group should be as large as possible, which means that a square must not be accounted for by itself if it can be accounted for by a group of 2 squares. A group of two squares should not be made if the involved squares can be included in a group of four squares and so on.

5. Don't care entries can be used in accounting for all of 1-squares to make optimum groups. They are marked 'X' in the corresponding squares. It is, however, not necessary to account for all don't care' entries.

Having accounted for groups with all 1s, the minimum sum-of-products or product-of-sums expressions can be written directly from the Karnaugh map. Min term Karnaugh map and Max term Karnaugh map of the Boolean function of a 2-input OR gate.

Min Term

A product term containing all the variables of the function in either complemented or un complemented form is called as min term.

Max Term

A sum term containing all the variables of the function in either complemented or un complemented form is called as max term.

Sum of Products (SOP)

A Sum-Of-Products Boolean expression is literally a set of Boolean terms added together each term being a multiplicative combination of Boolean variables.

Sum-of-products-expression = Term1 + Term2 ... + Term n

The product terms that include all of the input variables are called as min terms.

In a sum-of-products expression, we form a product of all the input variables (or their inverses) for each row of the truth table for which the result is logic 1. The output is the logical "sum" of these min terms.

The Sum-Of-Products expressions are easy to generate by determining which rows of the table have an output of 1, writing the one product term for each row and then finally sum all the product terms. This creates a Boolean expression which is representing the truth table as a whole. The Sum-Of-Products expressions lend themselves well to implementation as a set of AND gates (products) feeding into a single OR gate (sum).

Truth Table:

A	B	C	W
0	0	0	0
0	0	1	1
0	1	0	1
0	1	1	1
1	0	0	0
1	0	1	1
1	1	0	1
1	1	1	0

The output can be expressed as:

$$W=\overline{A}\bullet\overline{B}\bullet C+\overline{A}\bullet B\bullet\overline{C}+\overline{A}\bullet B\bullet C+A\bullet\overline{B}\bullet C+A\bullet B\bullet\overline{C}$$

Product of Sums (POS)

An alternative to a Sum-Of-Products expression to account for all "high" (1) output conditions in the truth table is to generate the Product-Of-Sums or POS expression to account for all "low" (0) output conditions instead.

POS Boolean expressions may be generated from the truth tables quite easily, by determining the rows of the table which have an output of 0, writing the one sum term for each row and then finally multiply all the sum terms. This creates a Boolean expression which is representing the truth table as a whole.

These "sum" terms that include all of the input variables (or their inverses) are called max terms. In the POS implementation, the output variable is the logical product of max terms. Product-Of-Sums expressions lend themselves well for implementation as a set of OR gates (sums) feeding into a single AND gate (product).

Truth table:

A	B	C	Z
0	0	0	1
0	0	1	1
0	1	0	1
0	1	1	1
1	0	0	0
1	0	1	0
1	1	0	0
1	1	1	1

The output can be expressed as:

$$Z=\left(\bar{A}+B+C\right)\cdot\left(\bar{A}+B+\bar{C}\right)\cdot\left(\bar{A}+\bar{B}+C\right)$$

Min term and Max term Boolean expressions for the two-input OR gate are as follows:

$Y = A+B$ (max term or product-of-sums).

$Y = \bar{A}\cdot B + A\cdot\bar{B} + A\cdot B$ (min term of sum-of -products).

Truth table:

A	B	Y
0	0	0
0	1	1
1	0	1
1	1	1

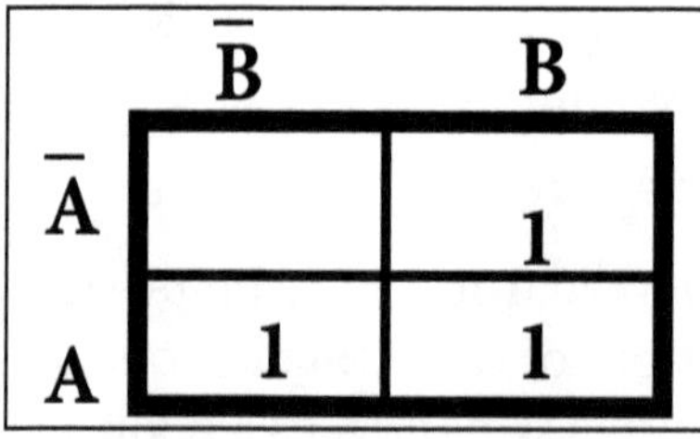

Sum-of-Products K-Map.

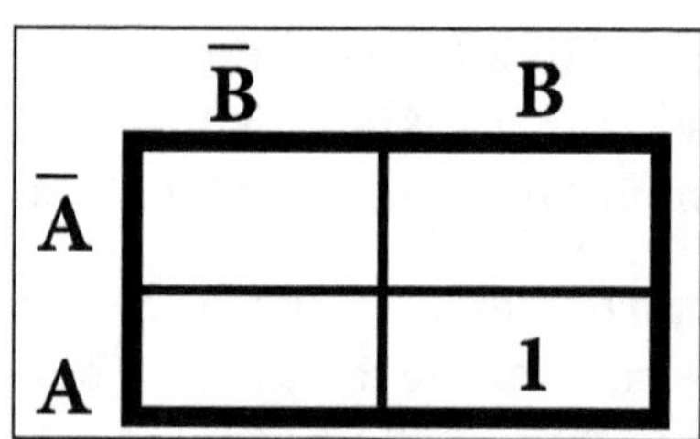

Product-of-Sums K-Map.

Min term Karnaugh Map and Max Term Karnaugh Map of the 3 Variable Boolean Functions

$$Y=\bar{A}\bullet\bar{B}\bullet\bar{C}+\bar{A}\bullet B\bullet\bar{C}+A\bullet\bar{B}\bullet\bar{C}+A\bullet B\bullet\bar{C}$$

$$Y=\left(\bar{A}+\bar{B}+\bar{C}\right)\cdot\left(\bar{A}+B+\bar{C}\right)\cdot\left(A+\bar{B}+\bar{C}\right)\cdot\left(A+B+\bar{C}\right)$$

Truth table:

A	B	C	Y
0	0	0	1
0	0	1	0
0	1	0	1
0	1	1	0
1	0	0	1
1	0	1	0
1	1	0	1
1	1	1	0

	$\bar{B}\bar{C}$	$\bar{B}C$	BC	$B\bar{C}$
$\bar{A}$	1			1
A	1			1

Sum-of-products K-map.

	$\bar{B}+\bar{C}$	$\bar{B}+C$	$B+C$	$B+\bar{C}$
$\bar{A}$	1			1
A	1			1

Products-of-sum K-map.

The Truth Table, Min Term Karnaugh Map and Max Term Karnaugh Map of the Four Variable Boolean Functions

$$Y=\bar{A}\cdot\bar{B}\cdot\bar{C}\cdot\bar{D}+\bar{A}\cdot\bar{B}\cdot\bar{C}\cdot D+\bar{A}\cdot B\cdot\bar{C}\cdot\bar{D}+\bar{A}\cdot B\cdot\bar{C}\cdot D+A\cdot\bar{B}\cdot\bar{C}\cdot\bar{D}$$

$$+A\cdot\bar{B}\cdot\bar{C}\cdot D+A\cdot B\cdot\bar{C}\cdot\bar{D}+A\cdot B\cdot\bar{C}\cdot D$$

$$Y=\left(A+B+\bar{C}+D\right)\cdot\left(A+B+\bar{C}+\bar{D}\right)\cdot\left(A+\bar{B}+\bar{C}+D\right)\cdot\left(A+\bar{B}+\bar{C}+\bar{D}\right)$$

$$\cdot\left(\bar{A}+B+\bar{C}+D\right)\cdot\left(\bar{A}+B+\bar{C}+D\right)\cdot\left(\bar{A}+\bar{B}+\bar{C}+D\right)\cdot\left(\bar{A}+\bar{B}+\bar{C}+\bar{D}\right)$$

To illustrate the process of forming groups and then writing the corresponding minimized Boolean expression, the below figures respectively shows min term and max term Karnaugh maps for the Boolean functions expressed by the below equations.

The minimized expressions as deduced from Karnaugh maps in the 2 cases are given by equation in the case of the min term Karnaugh map and equation in the case of the max term Karnaugh map:

$$Y=\bar{A}\cdot\bar{B}\cdot\bar{C}\cdot\bar{D}+\bar{A}\cdot\bar{B}\cdot C\cdot\bar{D}+\bar{A}\cdot B\cdot\bar{C}\cdot D+\bar{A}\cdot B\cdot C\cdot D+A\cdot\bar{B}\cdot\bar{C}\cdot\bar{D}$$

$$+A\cdot\bar{B}\cdot C\cdot\bar{D}+A\cdot B\cdot\bar{C}\cdot D+A\cdot B\cdot C\cdot D$$

$$Y=\left(A+B+C+\bar{D}\right)\cdot\left(A+B+\bar{C}+\bar{D}\right)\cdot\left(A+\bar{B}+C+D\right)\cdot$$

$$\left(A+\bar{B}+C+\bar{D}\right)\cdot\left(A+\bar{B}+\bar{C}+\bar{D}\right)$$

$$\cdot\left(A+\bar{B}+\bar{C}+D\right)\cdot\left(\bar{A}+\bar{B}+C+\bar{D}\right)\cdot\left(\bar{A}+\bar{B}+\bar{C}+\bar{D}\right)\cdot$$

$$\left(\bar{A}+B+C+\bar{D}\right)\cdot\left(\bar{A}+B+\bar{C}+\bar{D}\right)$$

$$Y=\bar{B}\cdot\bar{D}+B\cdot D$$

$$Y=\bar{D}\cdot\left(A+\bar{B}\right)$$

Truth Table:

A	B	C	D	Y
0	0	0	0	1
0	0	0	1	1
0	0	1	0	0
0	0	1	1	0
0	1	0	0	1
0	1	0	1	1
0	1	1	0	0
0	1	1	1	0
1	0	0	0	1
1	0	0	1	1
1	0	1	0	0
1	0	1	1	0
1	1	0	0	1
1	1	0	1	1

1	1	1	0	0
1	1	1	1	0

	$\bar{C}\bar{D}$	$\bar{C}D$	CD	$C\bar{D}$
$\bar{A}\bar{B}$	1	1		
$\bar{A}B$	1	1		
AB	1	1		
$A\bar{B}$	1	1		

Sum-of-products K-map.

	$\bar{C}+\bar{D}$	$\bar{C}+D$	$C+D$	$C+\bar{D}$
$\bar{A}+\bar{B}$	1	1		
$\bar{A}+B$	1	1		
$A+B$	1	1		
$A+\bar{B}$	1	1		

Products-of-sum K-map.

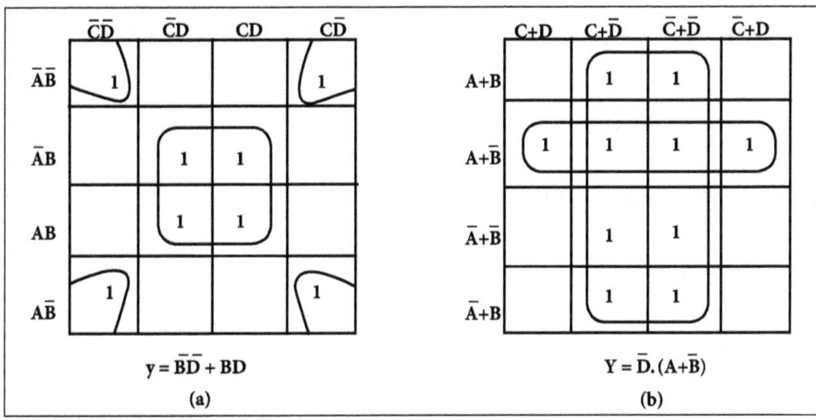

Group formation in min term and max term Karnaugh maps.

Problems

1. Let us show that a function $F = A + \bar{B}C$ expressed as a sum of its min terms is equivalent to a function expressed as a product of its maximum terms.

Solution:

Given:

$$F = A + \bar{B}C$$

POS:

$$F = A + \bar{B}C$$

$$=(A+\bar{B})(A+C)=(A+\bar{B}+C\bar{C})(A+C+B\bar{B})$$

$$=(A+\bar{B}+C)(A+\bar{B}+\bar{C})(A+B+C)$$

$$=M_2,M_3,M_0$$

$$F=\pi(0,2,3)$$

2. Let us express the function $F=A+\bar{B}C$ in Canonical SOP form.

Solution:

Given:

$$F=A+\bar{B}C$$

Sum of Product (SOP):

$$Y=A+\bar{B}\cdot C=A(B+\bar{B})+\bar{B}C(A+\bar{A})=AB+A\bar{B}+A\bar{B}C+\bar{A}\bar{B}C$$

$$=AB(C+\bar{C})+A\bar{B}(C+\bar{C})+A\bar{B}C+\bar{A}\bar{B}C$$

$$=ABC+AB\bar{C}+A\bar{B}C+A\bar{B}\bar{C}+\bar{A}\bar{B}C$$

$$Y=m_7+m_6+m_5+m_3+m_5+m_1$$

$$Y=\sum(1,4,5,6,7)$$

3. Here let us minimize and implement the following multiple output functions in SOP form.

$$f_1=\sum m\ (0,2,6,10,11,12,13)+d\ (3,4,5,14,15)$$

$$f_2=\sum m\ (1,2,6,7,8,13,14,15)+d\ (3,5,12)$$

Solution:

Given:

$$f_1=\sum m\ (0,2,6,10,11,12,13)+d\ (3,4,5,14,15)$$

$$f_2 = \sum m\ (1,2,6,7,8,13,14,15)\ +\ d\ (3,5,12)$$

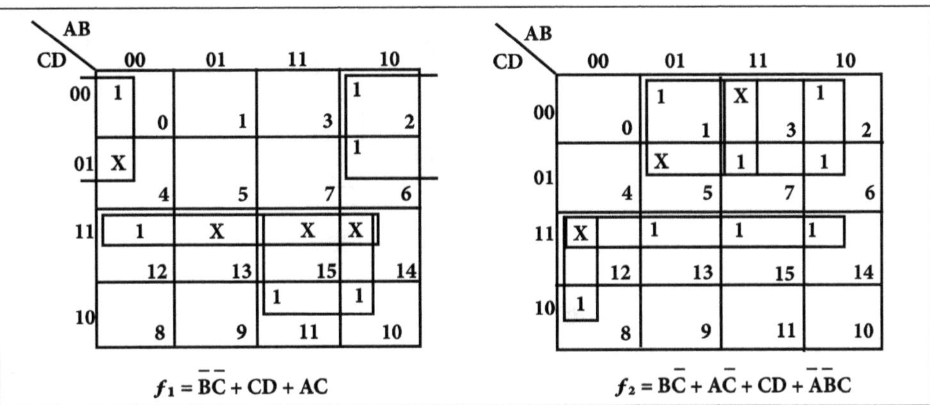

The K maps are filled ones and don't cares using the expression. After reduction, we find that CD occurs both in f_1 and f_2. So it can be shared.

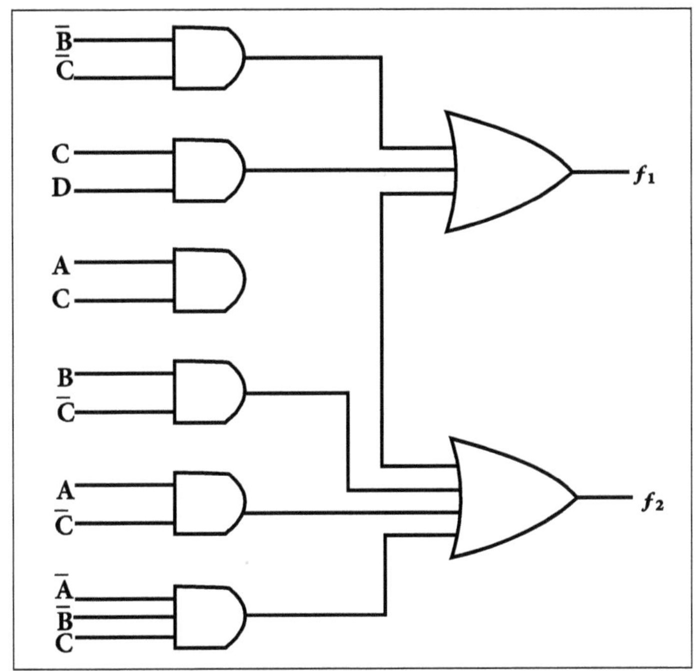

Logic Circuit Implementation.

4. Let us solve $g(W, X, Y, Z) = \sum m\ (1, 3, 4, 6, 11)\ +\ \sum d\ (0, 8, 10, 12, 13)$.

Solution:

Given:

$$g(W, X, Y, Z) = \sum m\ (1, 3, 4, 6, 11)\ +\ \sum d\ (0, 8, 10, 12, 13)$$

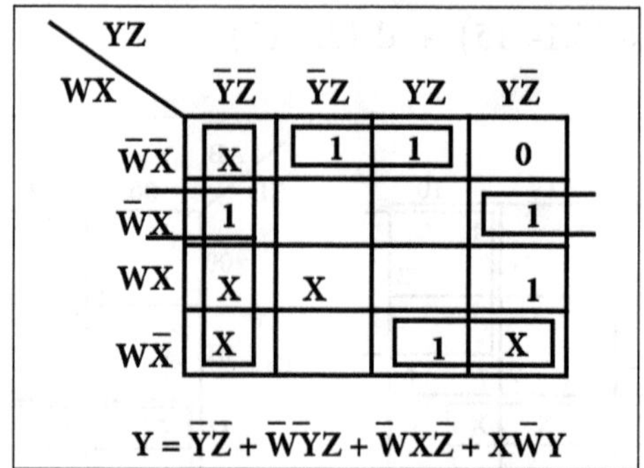

5. Let us simplify the following function using K - map and also implement the function using logic gates f(A, B, C) = π(0, 4, 6).

Solution:

Given:

$$f\,(A,\,B,\,C) = \pi\,(0,\,4,\,6)$$

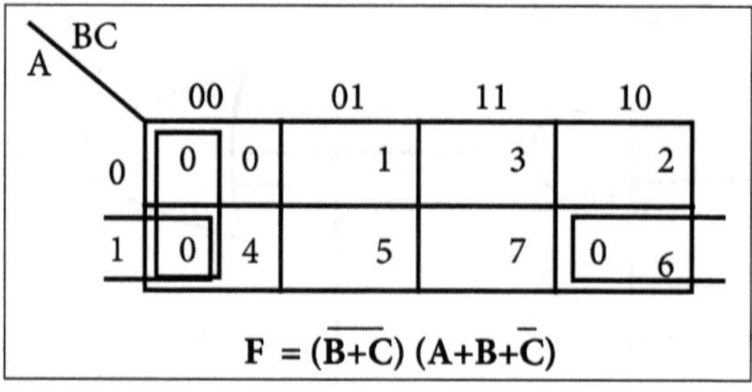

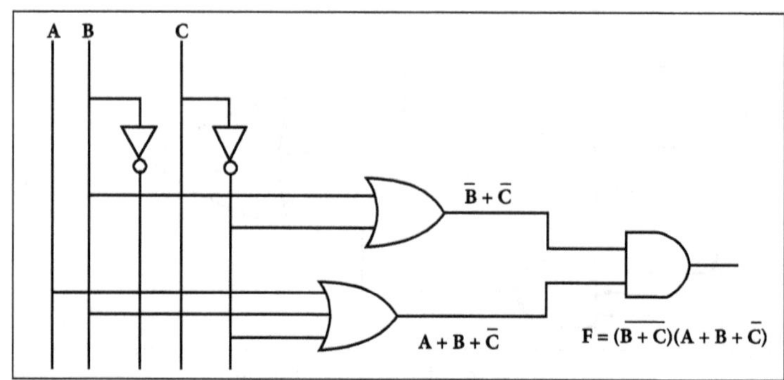

6. Let us reduce the following expression using K-map: F = m2 + m3 + m4 + m6 + m7 + m9 + m11 + m12.

Solution:

Given:

$$F = m2 + m3 + m4 + m6 + m7 + m9 + m11 + m12$$

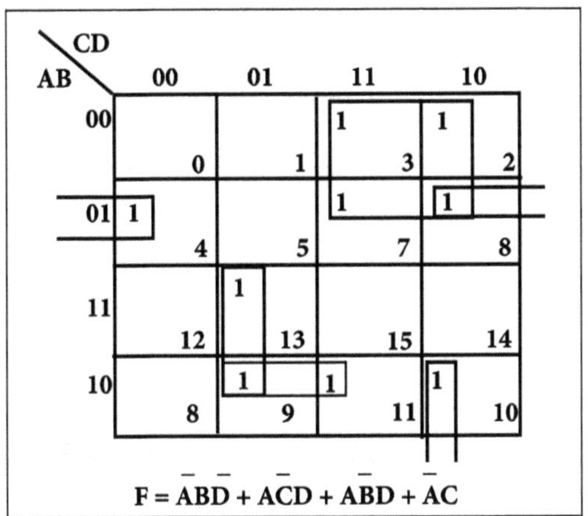

3.5.2 Design and Implementation of Digital Networks

The logic gates are the basic building blocks of any digital system. It is an electronic circuit having one or more than one input and only one output. Thus, the relationship between the input and the output is based on certain logic. Based on this, the logic gates are named as AND gate or gate, NOT gate etc.

Three Representations of Logic Functions

Logical functions	AND	OR	NOT
Expression	X.Y	X+Y	X =X'

AND Gate

The circuit which performs an AND operation is shown in figure. This gate has n input ($n \geq 2$) and one output.

Y = A AND B AND C.... N

Y = A.B.C........ N

Y = ABC .... N

The AND gate is so named because, if 0 is called "false" then 1 is called "true," the gate acts in the same way as the logical "and" operator. The output is "true" when both inputs are "true." Otherwise, the output is "false."

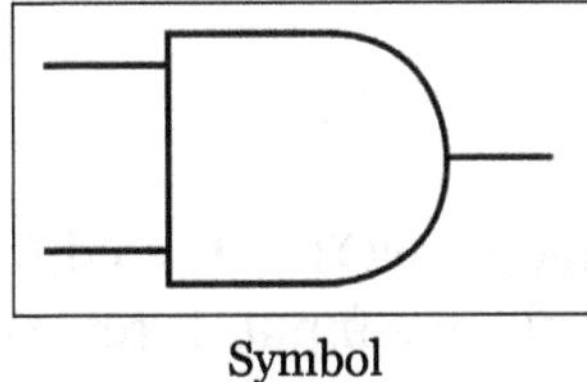

Symbol

Truth Table:

Input 1	Input 2	Output
0	0	0
0	1	0
1	0	0
1	1	1

OR Gate

The circuit which performs an OR operation is shown in figure. This gate has n input ($n \geq 2$) and one output.

Y = A OR B OR C .... N

$Y = A + B + C$ ...... N

The OR gate gets its name from the fact that it behaves like the logical inclusive "or." The output is "true" if either or both of the inputs are "true" and if both the inputs are "false," then the output is "false."

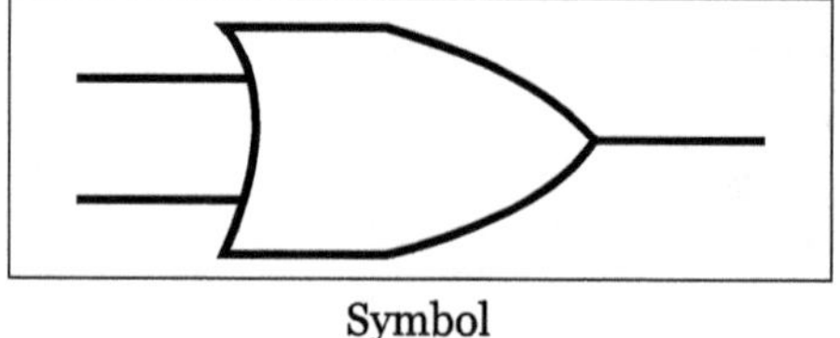

Symbol

Truth Table:

Input 1	Input 2	Output
0	0	0
0	1	1
1	0	1
1	1	1

NOT Gate

The NOT gate is also known as Inverter. This gate has one input A and one output Y.

Y = NOT A

$Y = \bar{A}$

A logical inverter, sometimes called a NOT gate to differentiate it from other types of electronic inverter devices, has only one input. It reverses the logic state.

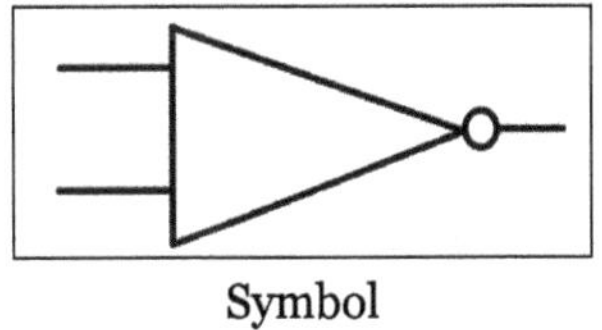

Symbol

Truth Table:

Input	Output
1	0
0	1

XOR Gate

The XOR or Ex-OR gate is a special type of gate. This gate can be used in the half adder, full adder and the sub tractor. Here, the exclusive-OR gate can be abbreviated as EX-OR gate or sometime as X-OR gate. This gate has n input (n ≥ 2) and one output.

$$Y = A \times OR\ B \times OR\ C \ .... \ N$$

$$Y = A \oplus B \oplus C \ .... \ N$$

The XOR (exclusive-OR) gate acts in the same way as the logical "either/or." The output is "true" if either, but not both the inputs are "true." The output is "false" when both the inputs are "true" or if both the inputs are "false".

Another way of analyzing this circuit is that the output is 1 when the inputs are different but 0 if the inputs are same.

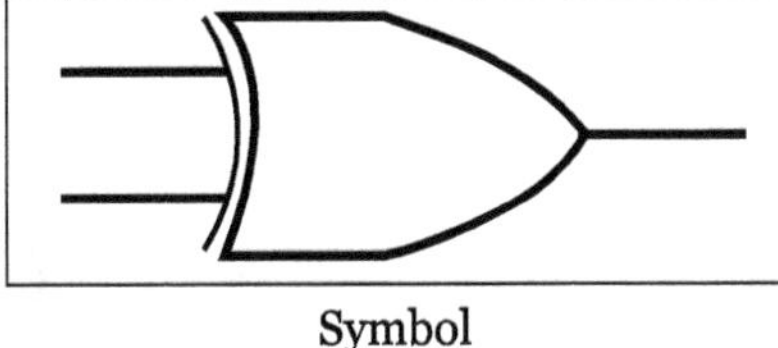

Symbol

Truth Table:

Input 1	Input 2	Output
0	0	0
0	1	1
1	0	1
1	1	0

XNOR Gate

The XNOR gate is a special type of gate. It can be used in the half adder, full adder and sub tractor. The exclusive-NOR gate is shortly known as EX-NOR gate or sometime as X-NOR gate. It has n input (n ≥ 2) and one output.

$$Y = A \times NOR\ B \times NOR\ C \ldots\ldots. N$$

The XNOR (exclusive-NOR) gate is a combination XOR gate followed by an inverter. Its output is "true" if the inputs are the same and "false" if the inputs are different.

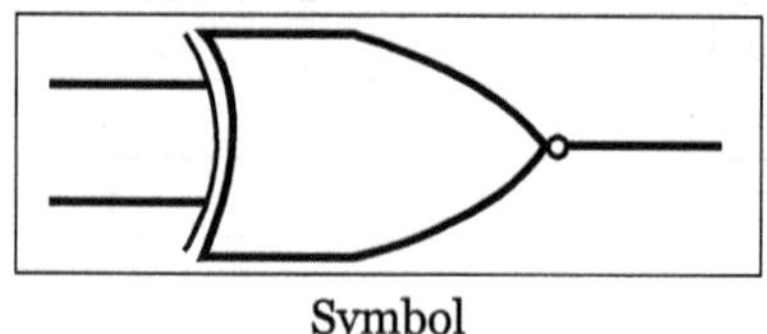

Symbol

Truth Table:

Input 1	Input 2	Output
0	0	1
0	1	0
1	0	0
1	1	1

Universal Gates - NAND, NOR Gates

NAND Gate

A NOT-AND operation is known as NAND operation. This gate has n input (n >= 2) and one output.

Y = A NOT AND B NOT AND C .... N

Y = A NAND B NAND C .... N

The NAND gate operates as an AND gate followed by a NOT gate. It acts in the similar manner of the logical operation "and", followed by negation. The output is "false" if both the inputs are "true". Otherwise, the output is "true."

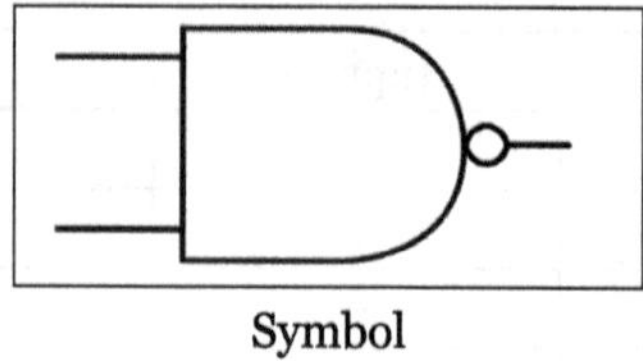

Symbol

Truth Table:

Input 1	Input 2	Output
0	0	1

0	1	1
1	0	1
1	1	0

NOR Gate

The NOT-OR operation is known as NOR operation. This gate has n input (n >= 2) and one output.

Y = A NOT OR B NOT OR C .... N

Y   A NOR B NOR C .... N

The NOR gate is a combination of OR gate followed by an inverter. Its output is “true” if both the inputs are “false”. Otherwise, the output is “false”.

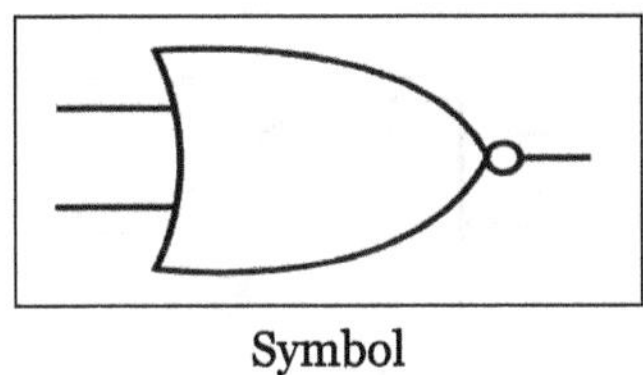

Symbol

Truth Table:

Input 1	Input 2	Output
0	0	1
0	1	0
1	0	0
1	1	0

3.5.3 Switching Circuits

The NAND and the NOR gates are generally faster and use fewer components than AND or OR gates. It can be implemented by using only NAND gates or only NOR gates with any logic function. NAND gate or NOR gate forms a functionally complete set where any switching function can be expressed in terms of NAND gate or NOR gate. Similarly, the set AND, OR and NOT shall form a functionally complete set since any Boolean function can be expressed in terms of SOP and POS only by using the AND, OR and NOT operations.

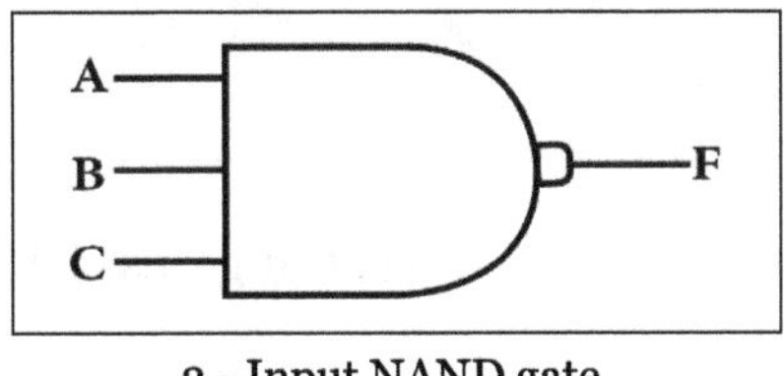

3 - Input NAND gate.

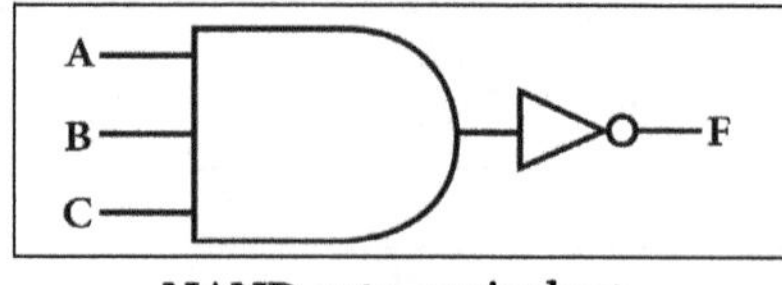

NAND gate equivalent.

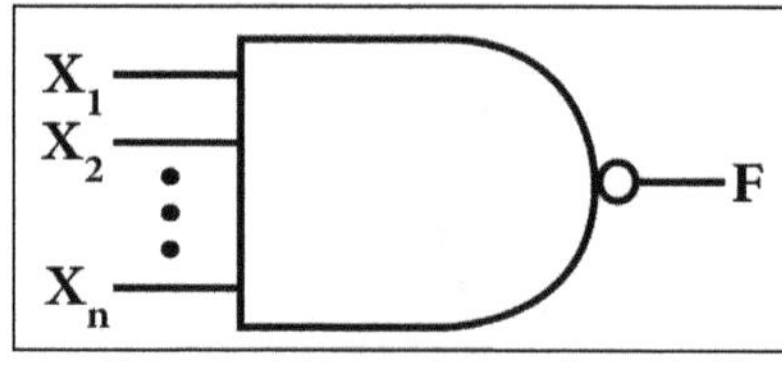

n-input NAND gate.

n -input NAND Gate

$$F = (X_1X_2.....X_n)' = X_1' + X_2' + ... + X_n'$$

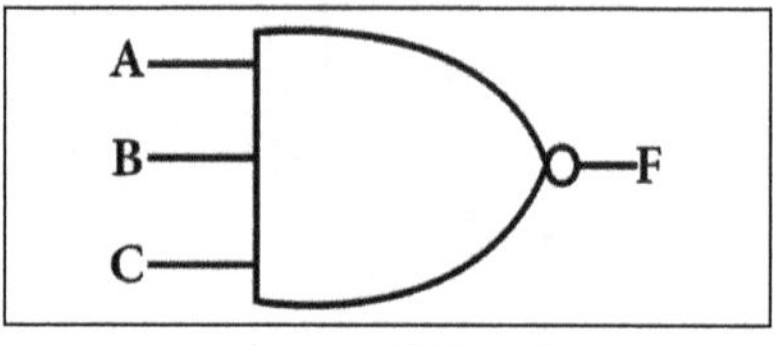

3 - Input NOR gate.

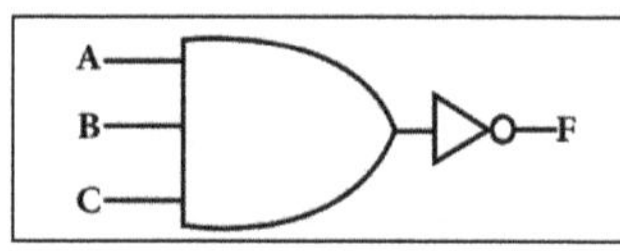

NOR gate equivalent.

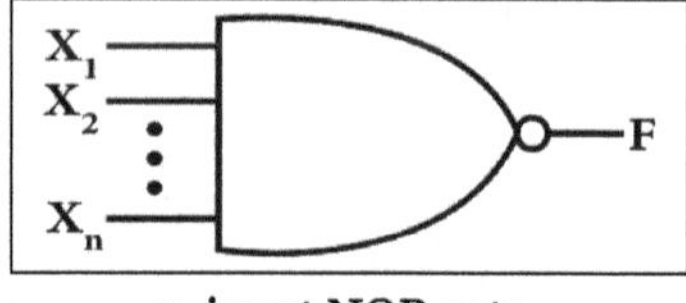

n-input NOR gate.

n- input NOR Gate

$$F = (X_1 + X_2 + ... + X_n)' = X_1'X_2'........ X_n'$$

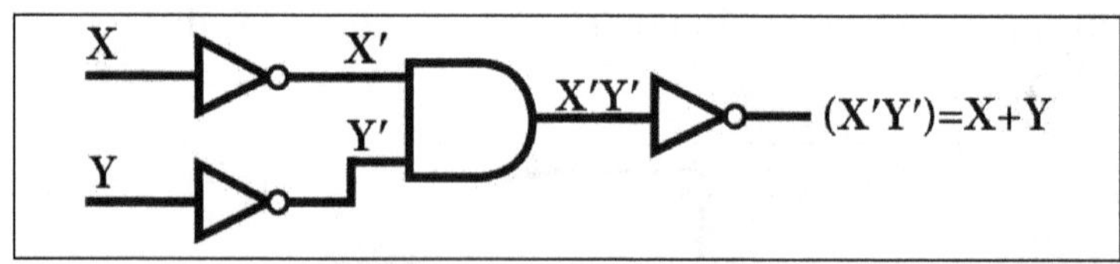

AND and NOT are a functionally complete set of gates. OR can be realized using AND and NOT.

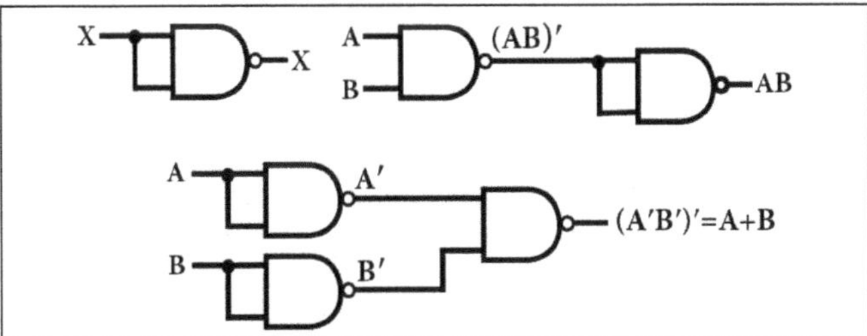

NAND Gate Realization of NOT and and OR.

Design of 2-Level Circuits using NAND -NAND Gates:

- Find a minimum SOP expression for F.
- Then, draw the corresponding 2-level AND-OR circuit.
- Replace all the gates with NAND gates leaving the gate interconnections unchanged. When the output gate has any single literals as inputs, complement these literals.

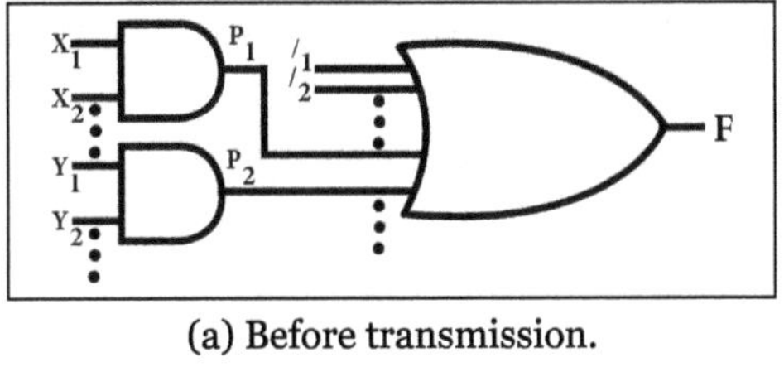

(a) Before transmission.

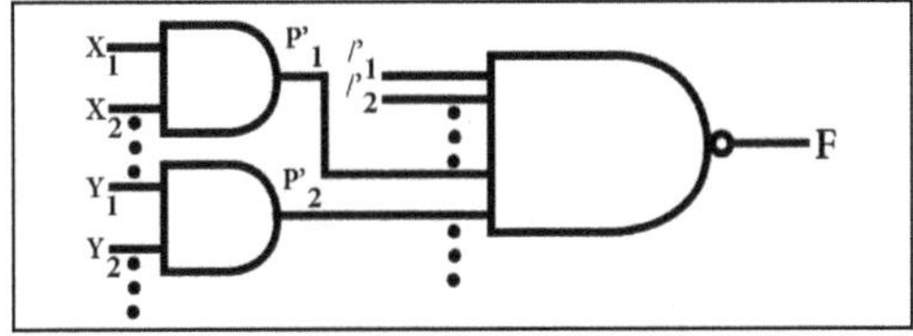

(b) After transmission.
AND-OR to NAND-NAND Transformation.

Steps for designing a minimum 2-level NOR-NOR circuit:

- Find a minimum POS expression for F.
- Draw the corresponding 2-level OR-AND circuit.
- Then, replace all gates with NOR gates leaving the gate interconnections unchanged. When the output gate has any single literals as inputs and complements these literals.

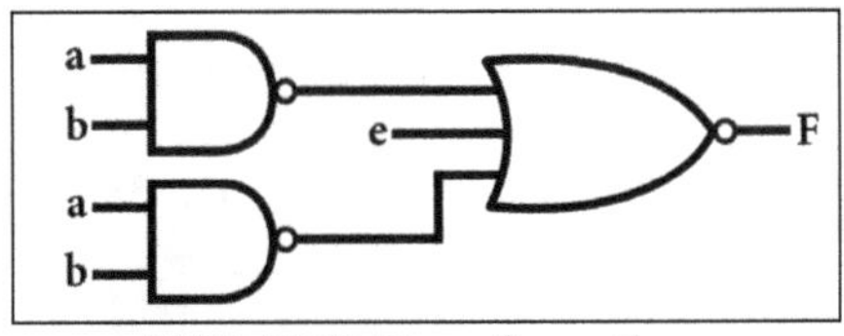

NOR-NOR Transformation.

$$F = \left[(ab)' + (cd)' + e\right]' = abcde'$$

Problems

1. Let us see the NAND gate implementation:

$$F(x, y, z) = (1,2,3,4,5,7)$$

Solution:

Given:

$$F(x, y, z) = (1,2,3,4,5,7)$$

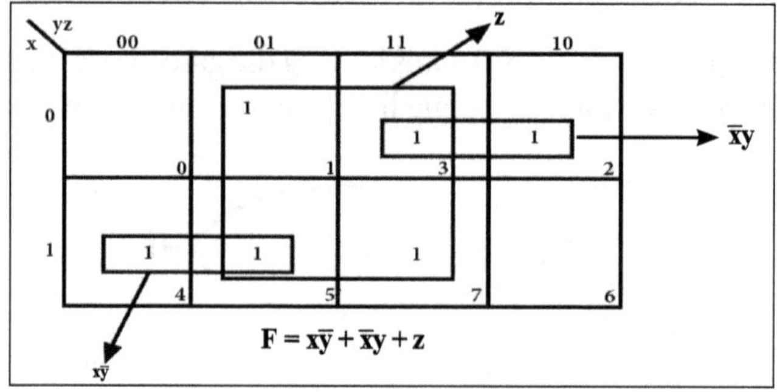

$$F = x\bar{y} + \bar{x}y + z$$

NAND Implementation

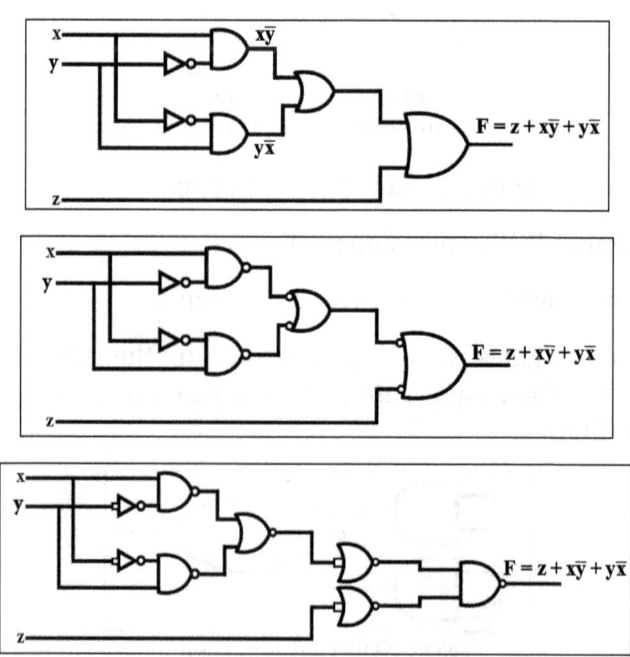

2. Let us draw the NAND logic diagram that implements the complement of the functions $F(A, B, C, D) = \sum(0, 1, 2, 3, 4, 8, 9, 12)$.

Solution:

Given:

$$F(A, B, C, D) = \sum(0, 1, 2, 3, 4, 8, 9, 12).$$

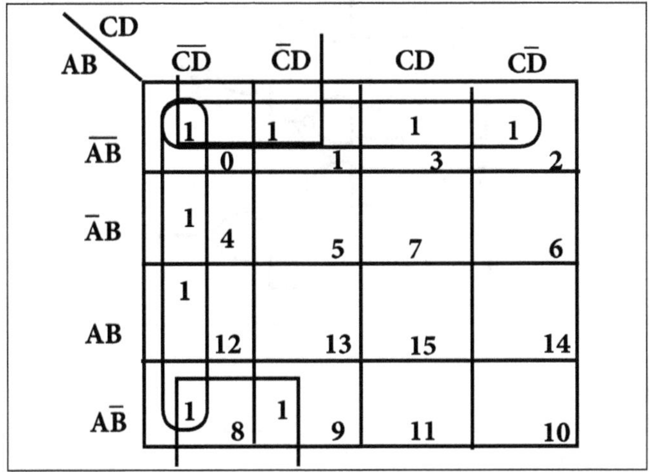

$$F = \bar{B}\bar{C}\bar{D} + \bar{B}\bar{C} + \bar{A}\bar{B}$$

$$F = \overline{\bar{C}\bar{D} + \bar{B}\bar{C} + \bar{A}\bar{B}}$$

$$= \left(\overline{\bar{C}\bar{D}}\right)\left(\overline{\bar{B}\bar{C}}\right)\left(\overline{\bar{A}\bar{B}}\right)$$

$$= (C + D)(B + C)(A + B)$$

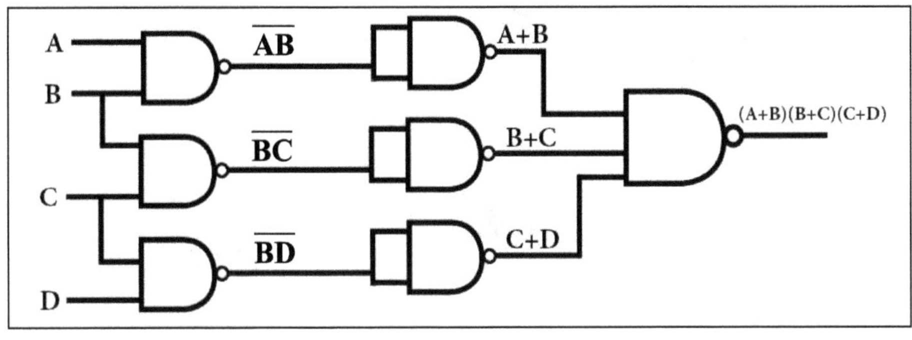

3. Let us see the implementation of EXOR using NAND and NOR gate.

Solution:

EXOR using NAND

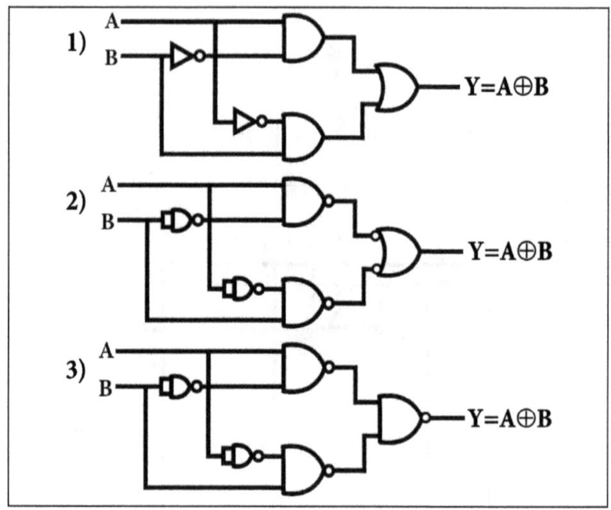

EXOR using NOR

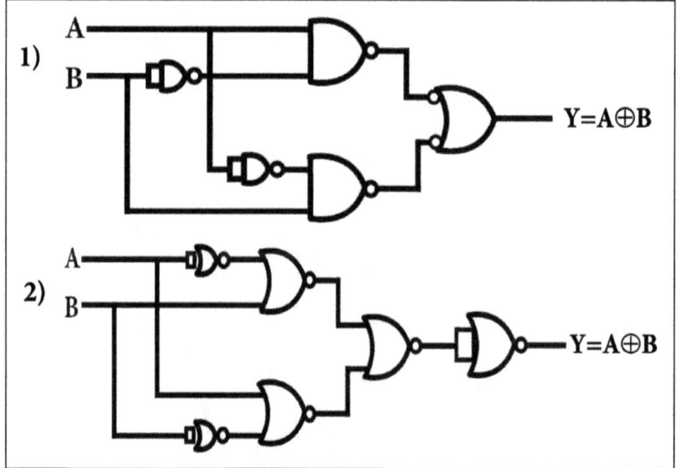

4

Graphs and Trees

4.1 Basic Terminology

Graph theory concerns the relationship among the lines and points. Graph consists of some points and some lines between them. No attention is paid to the position of points and the length of the lines.

A simple graph G is one that satisfies the following:

- Having at most one edge (line) between any two vertices (points).
- Not having an edge coming back to the original vertex.

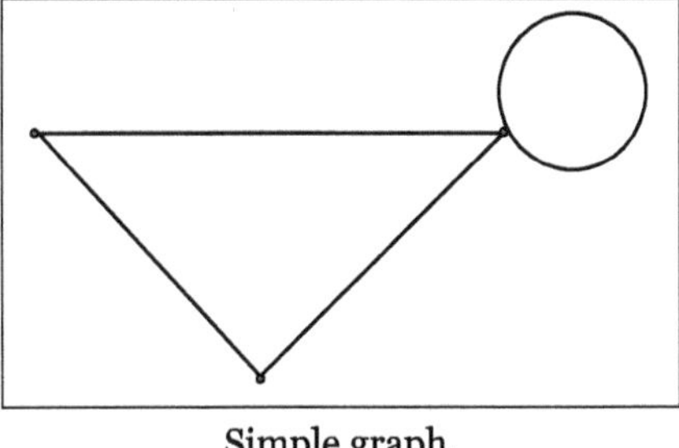

Simple graph.

A simple graph G consists of a non-empty finite set V (G) of elements called vertices (or nodes) and a finite set E (G) of distinct unordered pairs of distinct elements of V (G) called edges. We call V (G) the vertex set and E (G) the edge set of G. An edge {v, w} is said to join the vertices v and w and is usually abbreviated to vw. For example, figure (a) represents the simple graph G whose vertex set V (G) is {w, v, w, z} and whose edge set E (G) consists of the edges uv, uw, vw and wz.

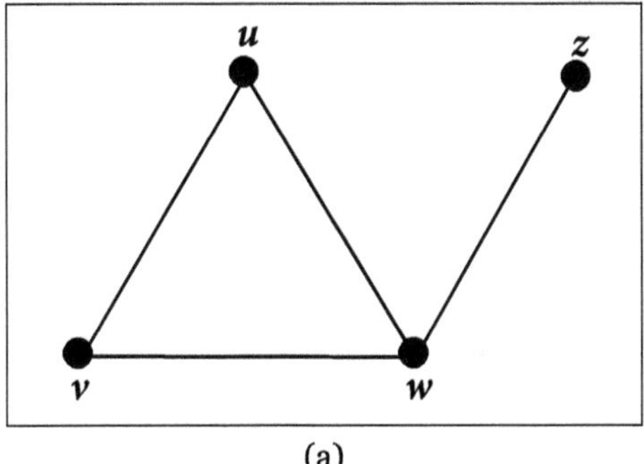

(a)

In any simple graph there is at most one edge joining a given pair of vertices. However, many results that hold for simple graphs can be extended to more general objects in which two vertices may have several edges joining them.

In addition, we may remove the restriction that an edge joins two distinct vertices allow loops - edges joining a vertex to it. The resulting object, in which loops and multiple edges are allowed, is called a general graph or simply a graph (figure (b)). Thus, every simple graph is a graph, but not every graph is a simple graph.

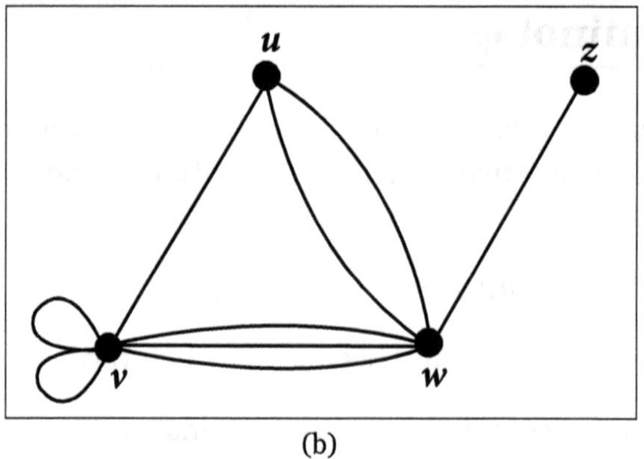

(b)

Thus, a graph G consists of a non-empty finite set V(G) of elements called vertices, a finite family E(G) of unordered pairs of elements of V(G) called edges.

We call V(G) as the vertex set and E(G) as the edge family of G. An edge {v, w} is said to join the vertices v and w, is again abbreviated to vw. Thus, in figure (b), V(G) is the set {u, v, w, z} and E(G) consists of the edges uv, vv (twice), vw (three times), uw (twice) and wz. Note that each loop vv joins the vertex v to itself.

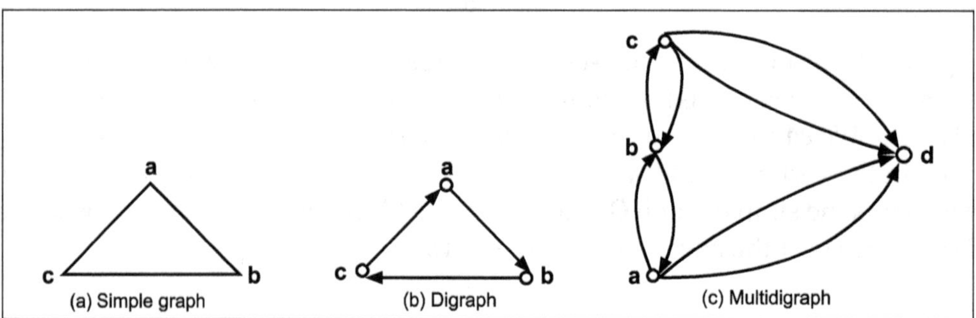

(a) Simple graph (b) Digraph (c) Multidigraph

Graph

A simple directed graph (or digraph) G = (V, E) consists of a non-empty set of vertices V and a set of directed edges (or arcs) E. Each directed edge is associated with an ordered pair of vertices. The directed edge associated with the ordered pair (u, v) is said to start at u and end at v.

Multi-graph:

A Multi graph G = (V, E) consists of a non-empty set V of vertices, a set E of edges made of pairs of distinct elements of V and a function f from E to {{u, v} | u, v ∈ V ∧u 6= v}. Edges e_1 and e_2 are called multiple edges (or parallel edges) if f (e_1) = f (e_2).

Branch:

A line segment replacing one or more network elements which are connected in series or parallel.

Node:

Interconnection of two or more branches. It is the terminal of the branch. Usually the interconnections of three or more branches are nodes.

Path:

A set of branches that may be traversed in the order without passing through the same node more than once.

Loop:

Any closed contour selected in the graph.

Mesh:

A loop which does not contain any other loop within it.

Planar graph:

A graph which may be drawn on the plane surface in such a way that no branch has passes over any other branch.

Non-planar Graph:

Any graph which is not planar.

Oriented Graph:

When a direction to each branch of the graph is assigned, the resulting graph is called as an oriented graph or a directed graph.

Connected Graph:

A graph is connected if and only if there is a path between every pair of nodes.

Sub Graph:

Any subset of the branches of the graph.

Tree

A connected sub-graph containing all the nodes of a graph but no closed path. i.e. it is a set of branches of graph which contains no loop but connects every node to every other node not necessarily directly. Number of different trees can be drawn for the given graph.

Link

A branch of the graph which does not belong to the particular tree under consideration. The links form the sub-graph not necessarily connected and is called as the co-tree.

Tree Compliment:

Totality of links i.e. co-tree.

Independent Loop

The addition of each link to a tree, one at a time, results one closed path called an independent loop. Such a loop contains only one link and other tree branches. Obviously, the number of such independent loops equals the number of links.

Tie Set

A set of branches contained in the loop such that each loop contains one link and the remainder are the tree branches.

Tree Branch Voltages

The branch voltages may be separated into tree branch voltages and link voltages. The tree branches connect all the nodes. Thus, if the tree branch voltages are forced to be zero, then all the node potentials become coincident and thus all the branch voltages are forced to be zero. As the act of setting only the tree branch voltages to zero forces all voltages in the network to be zero, it must be possible to express all the link voltages uniquely in terms of the tree branch voltages. Thus, the tree branch form an independent set of equations.

Cut Set

A set of elements of the graph that dissociates it into two main portions of a network such that replacing any one element will destroy this same property. It is a set of branches that if removed divides the connected graph into two connected sub-graphs. Each cut set contains one tree branch and the remaining being links. Figure shows a typical network with its graph oriented graph, a tree, co-tree and a non-planar graph.

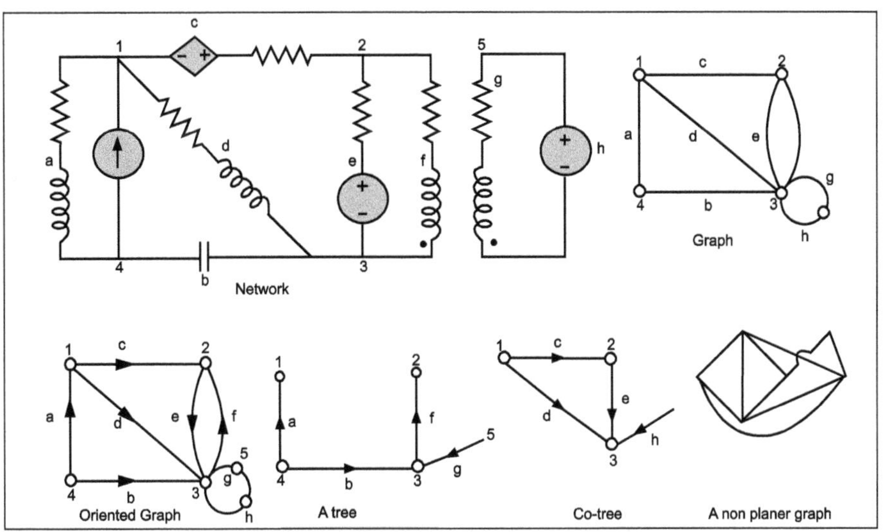

Relation between Nodes, Links and Branches

Let B = Total number of branches in the graph or network

N = total nodes

L = link branches

Then, N−1 branches are required to construct the tree because the first branch chosen connects the two nodes and each additional branch includes one more node.

Therefore, number of independent node pair voltages =N−1= number of tree branches.

Then, L=B− (N−1) =B−N+1

Number of independent loops =B−N+1

Isomorphic Graphs

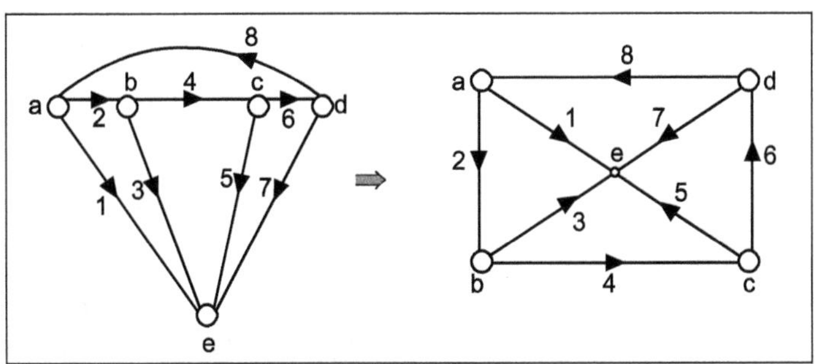

Isomorphic graphs.

Two graphs are said to be isomorphic if they have the same incidence matrix, though they look different. It means that they have the same numbers of nodes and the same numbers of branches. There is one to one correspondence between the nodes and one to one correspondence between the branches.

Problems

1. If R = (x/y) / x > y is a relation on X = {1, 2, 3, 4}, let us draw the graph of R.

Solution:

Given:

$$R = (x/y)/\ x > y \text{ is a relation on } X = \{1,\ 2,\ 3,\ 4\}$$

$$R = \{(x,\ y)/x\ >\ y\}$$

$$R = \{(2,1),\ (3,1),\ (4,1),\ (3,2),\ (4,2),\ (4,3)\}.$$

The graph of R:

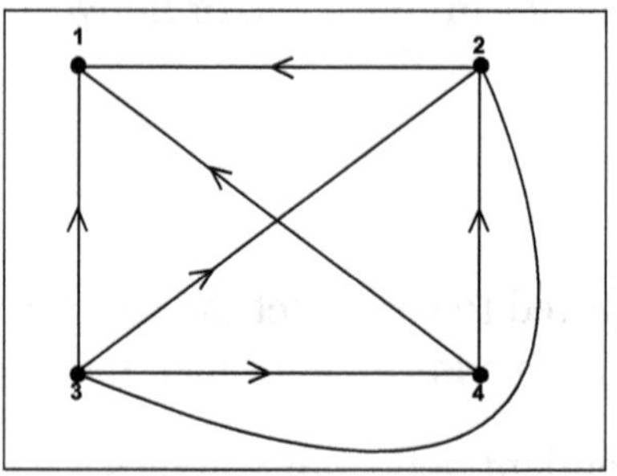

Graph terminology.

2. Let us find the equivalence relation on the set S = {1, 2, 3, 4, 5} which generates the partition $\{\overline{1,2}, \overline{3}, \overline{4,5}\}$. Let us also draw the graph of the relation.

Solution:

Given:

$$S = \{1,\ 2,\ 3,4,\ 5\}$$

Let P be a partition on $S = \{\overline{1,2}, \overline{3}, \overline{4,5}\}$

Let R be the relation determine by the portion P.

{1, 2} is a block, the elements 1, 2 are related only to the elements of the subset {1, 2}.

∴ (1, 1), (1, 2), (2, 1), (2, 2) ∈ R.

Similarly the block 3 so (3, 3) ∈ R.

The block {4, 5}, so (4, 4), (4, 5), (5, 4), (5, 5) ∈ R.

Hence, the relation R = {(1, 1), (1, 2), (2, 1), (2, 2), (3, 3), (4, 4), (4, 5), (5, 4), (5, 5)}. The graph of the relation:

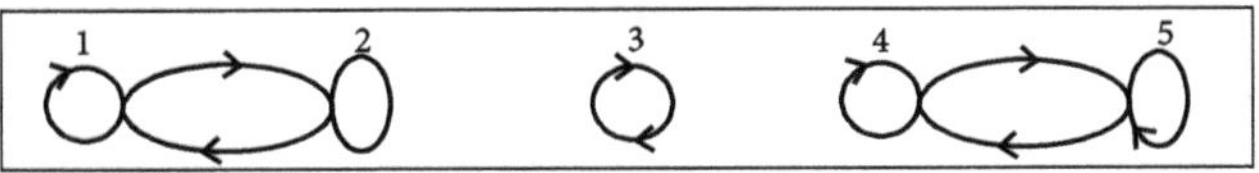

3. Let us draw a complete bipartite graph of $K_{2,3}$ and $K_{3,3}$.

Solution:

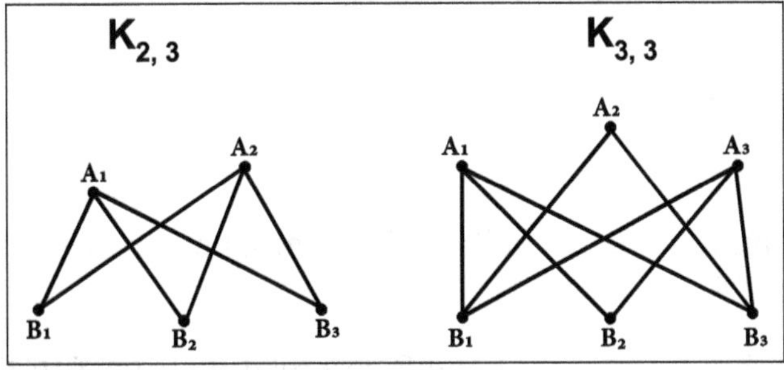

4. Let us find the connected sub graph obtained from the graph given in the following figure, by deleting each vertex. Let us list out the simple paths from A to F in each sub graph.

Solution:

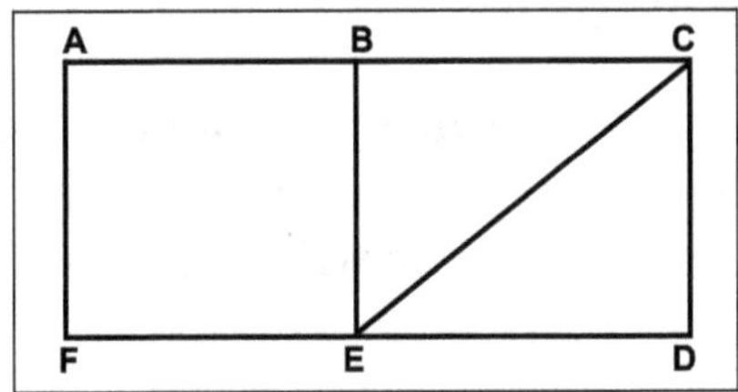

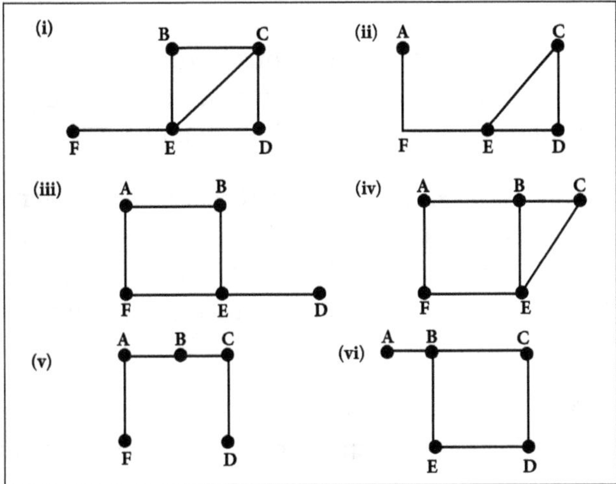

Simple path from A to F

Figure (i) No path from A to F

Figure (ii) A - F

Figure (iii) A - F, A - B - E - F

Figure (iv) A - F, A - B - E - F; A - B - C - E - F

Figure (v) A - F;

Figure (vi) No path from A to F (No. F Vertex).

4.1.1 Digraphs and Relations

Relation Matrix

A relation R from a finite set X to a finite set Y can also be represented by a matrix called the relation matrix of R.

Let X = {x_1, x_2... x_m}, Y = {y_1, y_2, ..., y_n} and R be a relation from X to Y. The relation matrix of R can be obtained by first constructing a table whose columns are preceded by a column consisting of successive elements of X and whose rows are headed by a row consisting of the successive elements of Y.

Digraphs and Binary Relations

Let A and B be the non-empty sets. A (binary) relation R from A to B is a subset of A × B. If R ⊆ A x B and (a, b) ∈ R, where a ∈ A, b ∈ B, we say a "is related to" b by R and we write a R b. If a is not related to b by R, we write a o b. A relation R defined on a set X is the subset of X × X.

Example:

Less than, greater than and equality are relations in the set of real numbers. The property "is congruent to" defines the relation in the set of all triangles in a plane. Also, parallelism defines a relation in the set of all lines in a plane.

Let R define a relation on a non-empty set X. If R relates every element of X in to itself, the relation R is reflexive. A relation R is said to be symmetric if for all x_i x_j ∈ X, x_i R x_j means x_j R x_i. A relation R is transitive if for any three elements x_i, x_j and x_k in X, x, Rx and x_j R x_k means x_i/2 x_k. A binary relation is known as an equivalence relation if it is reflexive, symmetric and transitive.

A binary relation R on a set X may always be represented by a digraph. In such a representation, each x_j ∈ X is represented by a vertex x_i and whenever there is a relation R from x_i to x_j, an arc is drawn from x_i to x_j, for every pair (x_i, x_j). The digraph in the figure represents the relation is less than, on a set consisting of four numbers 2, 3, 4, 6.

Every binary relation on a finite set is represented by a digraph without the parallel edges and vice-versa. Clearly, digraph of reflexive relation contains a loop at every vertex. A digraph representing a reflexive binary relation is known as the reflexive digraph.

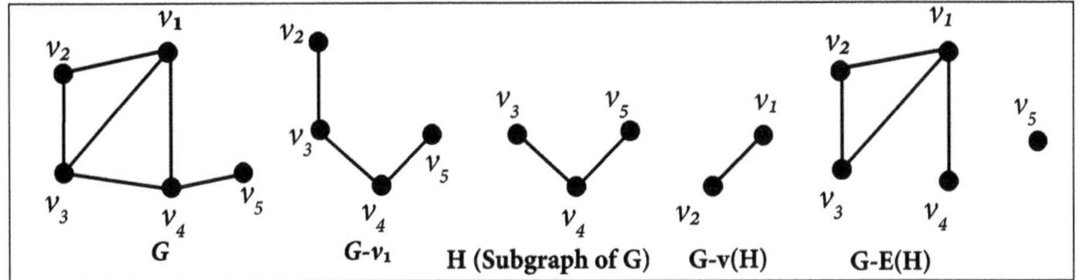

Digraph of reflexive relation.

The digraph of a symmetric relation is a symmetric digraph since for every arc from x_i to x_j, there is an arc from x_j to x_i. Figure shows the digraph of an irreflexive and the symmetric relation on a set of three elements.

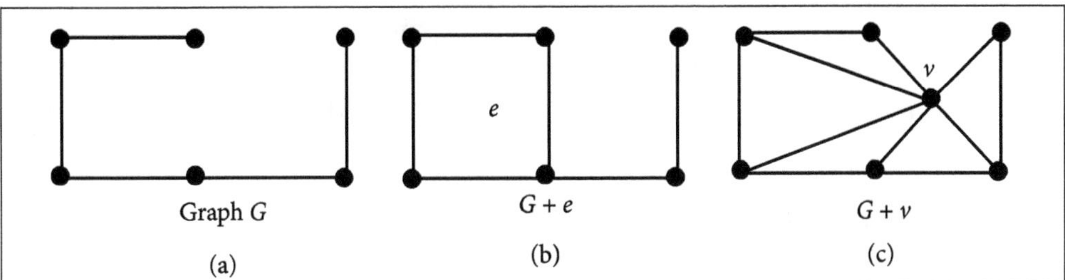

Digraph of an irreflexive and the symmetric relation on a set of three elements.

A digraph representing a transitive relation on its vertex set is known as a transitive digraph. Figure below explains the digraph of a transitive that is neither reflexive, nor symmetric.

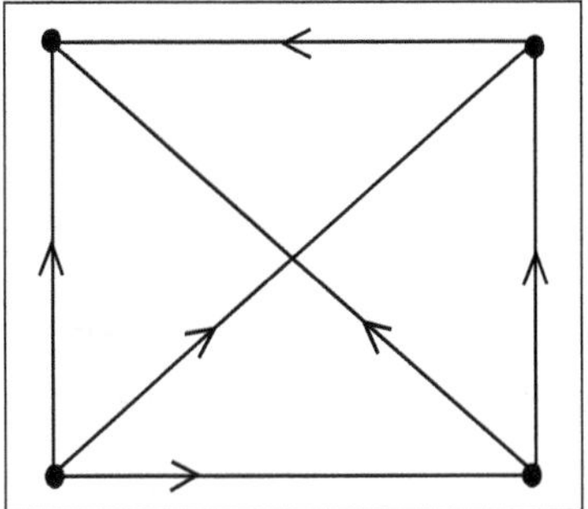

Digraph of a transitive relation.

A binary relation R on a set M can also be represented by a matrix, known as a relation matrix. This is a (0, 1), n × n matrix $M_R = [m_{i\,j}]$, where n is the number of elements in M and is defined by

$$m_{ij} = \begin{cases} 1 & \text{if } x_i \text{ R } x_j \text{ is true,} \\ 0, & \text{otherwise.} \end{cases}$$

The relation between the objects is not symmetric in some problems. For these cases we require directed graphs, where the edges are oriented from one vertex to another.

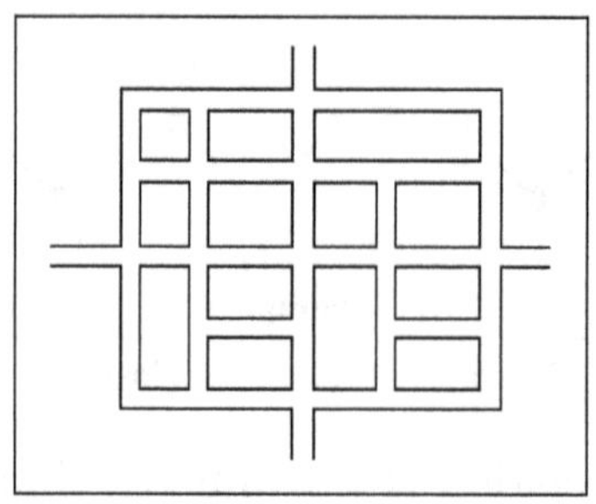

Digraph

A digraph $D = (V_D, E_D)$ consists of the vertices V_D and (directed) edges $E_D \subseteq V_D \times V_D$ (without loops vv). We still write uv for (u, v), but note that now uv ≠ vu. For each pair, e = uv define the inverse of e as e^{-1} = vu (= (v, u)). Note that e ∈ D does not imply e^{-1} ∈ D.

Let D be a digraph. Then A is further classified into:

Sub digraph, if $V_A \subseteq V_D$ and $E_A \subseteq E_D$,

Induced sub digraph, A = D[X], if VA = X and $E_A = E_D \cap (X \times X)$.

The underlying graph U (D) of a digraph D is the graph on V_D such that if e ∈ D, then the undirected edge with same ends are in U (D).

A digraph D is an orientation of a graph G, if G = U (D) and e ∈ D implies e^{-1} /∈ D.

In this case, D is said to be an oriented graph.

Example:

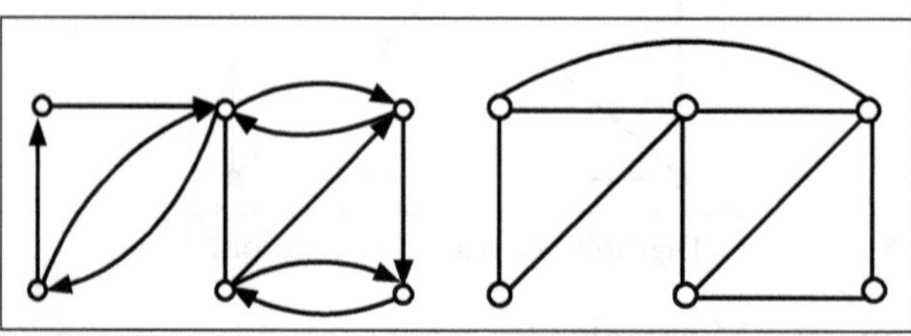

4.1.2 Representation of Graphs

One way to represent a graph without multiple edges is to list all the edges of this graph.

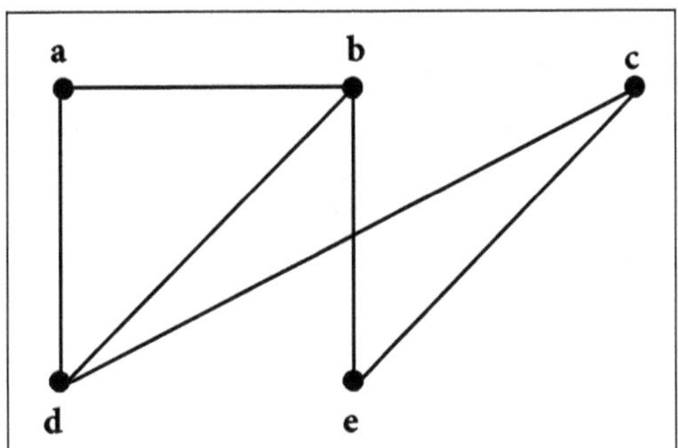

$G = (V, E)$ with $V = \{a, b, c, d, e\}$ and $E = \{\{a, b\},$

$\{a, d\}, \{b, d\}, \{b, e\}, \{d, c\}, \{e, c\}\}.$

Representing Graphs by Adjacency Lists

Another way to represent a graph without multiple edges is to use adjacency lists, which specify the vertices that are adjacent to each vertex of the graph.

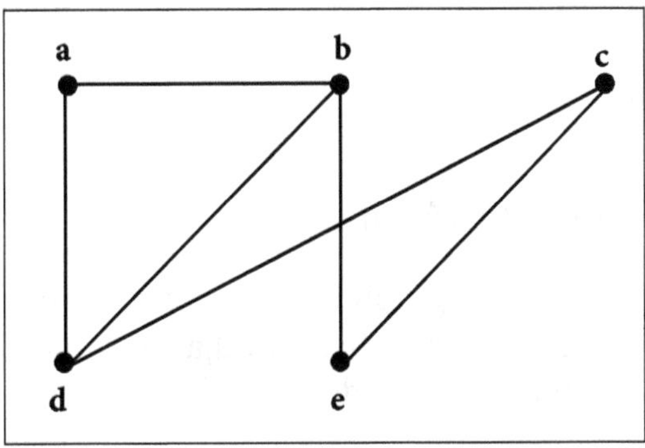

Vertex	Adjacent vertices
A	b, d
b	a, c, d, e
c	d, e
d	a, b, c
e	b, c

Adjacency Matrices

Definition

Suppose that $G = (V, E)$ is a simple graph where $|V| = n$. Suppose that the vertices of G are listed arbitrarily $v_1, v_2, \ldots, v_n$.

The adjacency matrix A (or AG) of G, with respect to this listing of the vertices, is the n × n zero-one matrix with 1 as its (i , j)th entry when v_i and v_j are adjacent and 0 as its (i ,j) th entry when they are not adjacent.

Example of Adjacency Matrices

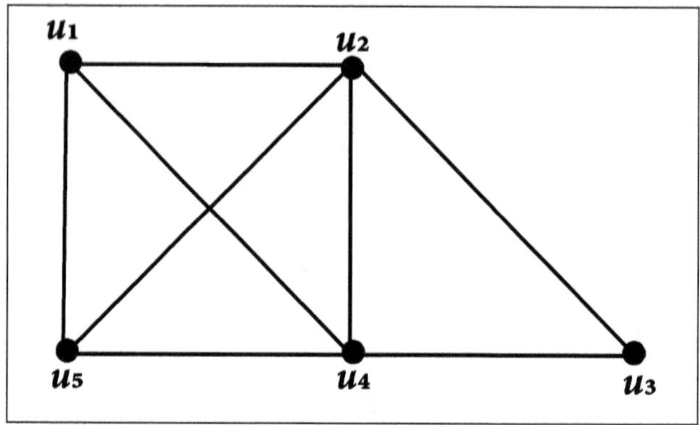

We order the vertices as u_1, u_2, u_3, u_4, u_5.

$$\begin{pmatrix} 0 & 1 & 0 & 1 & 1 \\ 1 & 0 & 1 & 1 & 1 \\ 0 & 1 & 0 & 1 & 0 \\ 1 & 1 & 1 & 0 & 1 \\ 1 & 1 & 0 & 1 & 1 \end{pmatrix}$$

Remarks on Adjacency Matrices

Note that an adjacency matrix of a graph is based on the ordering chosen for the vertices. Hence, there are as many as n! different adjacency matrices for a graph with n vertices, because there are n! different orderings of n vertices.

The adjacency matrices of a simple graph is symmetric because if v_i is adjacent to v_j, then v_j is adjacent to v_i and if v_i is not adjacent to v_j, then v_j is not adjacent to v_i. Since a simple graph cannot have a loop, $a_i = 0$ for i = 1, 2, ..., n.

Adjacency Matrices for Pseudo Graphs

A loop on the vertex vi is denoted by a 1 at the (i, i)th position of the adjacency matrix.

When there are multiple edges between two vertices, the (i , j)th element of the adjacency matrix is equal to the number of edges between vertices v_i and v_j.

All undirected graphs, including simple graphs, multi graphs and pseudo graphs, have symmetric adjacency matrices.

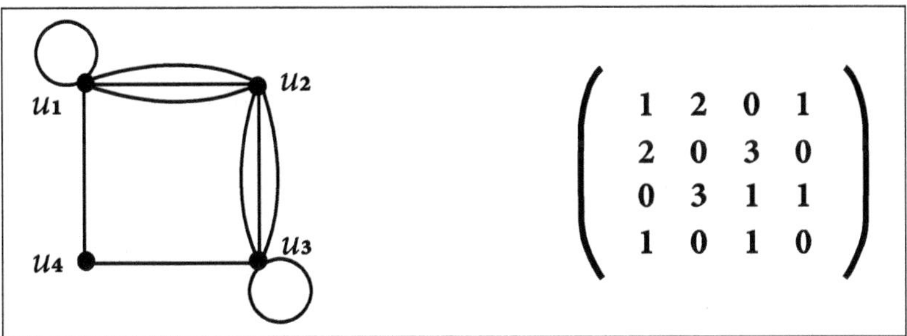

4.1.3 Operations on Graphs

If G = (V, E) and G* = (V*, E) are any two graphs then union, intersection, ring sum etc. operations can be defined on graphs.

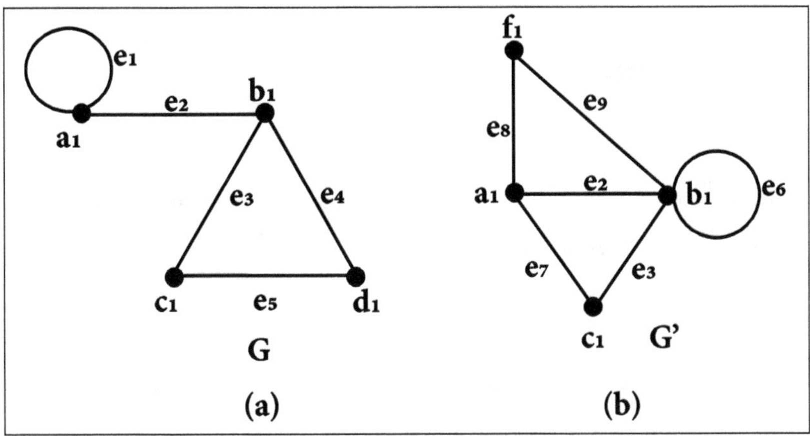

Where,

$$V = \{a_1, b_1, c_1, d_1\}$$

$$E = \{e_1, e_2, e_3, e_4, e_5\}$$

$$V^* = \{a_1, b_1, c_1, f_1\}$$

$$E^* = \{e_2, e_3, e_6, e_7, e_8, e_9\}$$

Union of Graphs

For the graphs G and G*

Union G∪G* is a graph whose vertex set is V∪V* and edge set is E∪E*

Where G = (V, E) and G* = (V*, E*)

For the above figure:

G∪G* is a graph with vertex set{a_1, b_1, c_1, d_1, f_1} and edge set is {e_1, e_2, e_3, e_4, e_5, e_6, e_7, e_8, e_9}

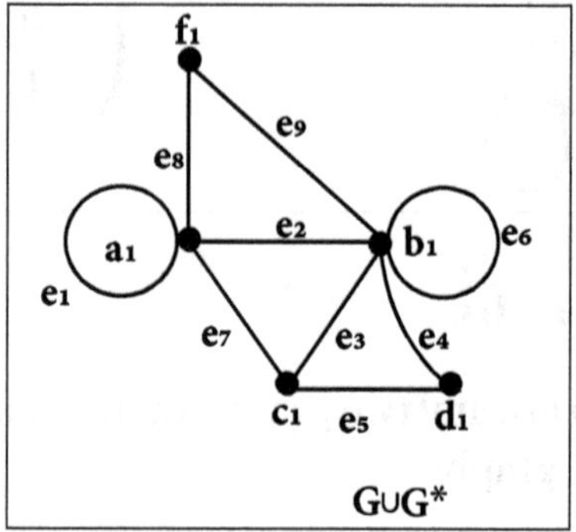

G∪G*

Intersection of Graphs

For the graphs G and G*.

Intersection G∩G* is a graph whose vertex set is V∩V* and edge set is E∩E*

Where G = (V, E) and G* = (V*, E*)

For the above figure:

G∩G* is a graph with vertex set {a_1, b_1, c_1} and edge set is {e_2, e_3}

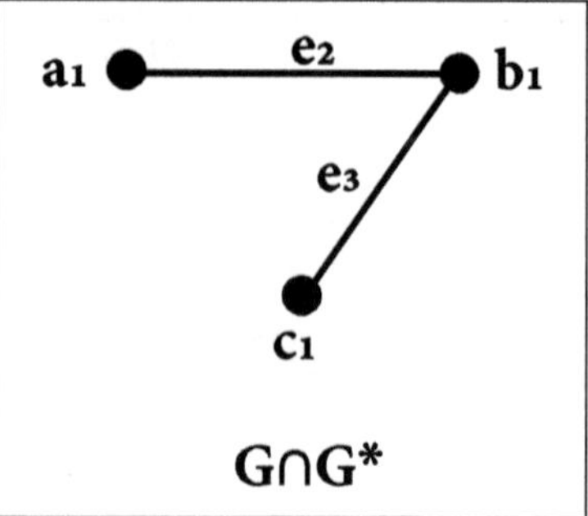

G∩G*

Ring Sum of Graphs

For the graphs G and G*.

Ring sum G ⊕ G* is a graph whose vertex set is V∪V* and edge set is a set of those edges which are either in G or G* but not in both.

For the graphs shown in the above figure (a) and (b),

G⊕G* is a graph with vertex set {a_1,b_1,c_1,d_1, f_1} and edge set is {e_1,e_4,e_5,e_6,e_7,e_8,e_9}

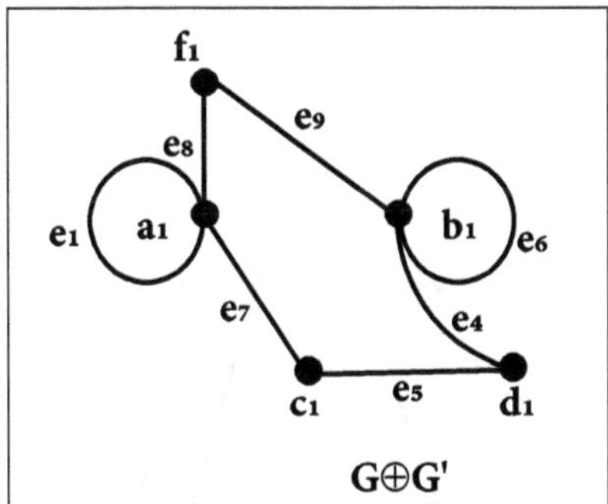

Removal of an Edge

For a graph G = (V, E).

Let e ∈ E, then the graph G - e is the graph obtained by removing an edge e from the graph.

Example:

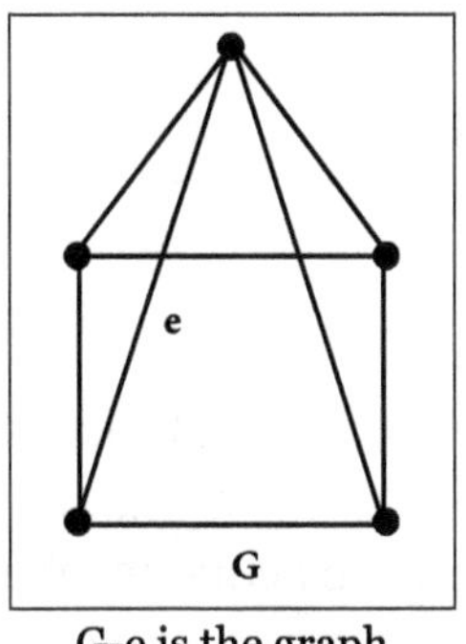

G-e is the graph

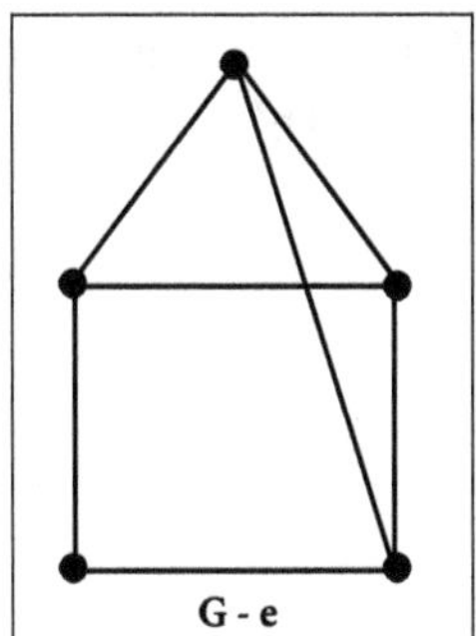

Removal of a Vertex

For a graph G = (V, E).

Let v ∈ V , then the graph G - v is obtained by removing the vertex v from the graph G. Removal of v means, removal of all these edges also which are incident on V.

Example:

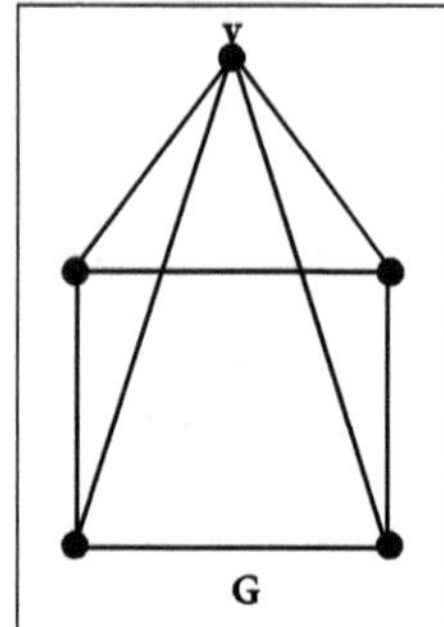

G-v is the graph.

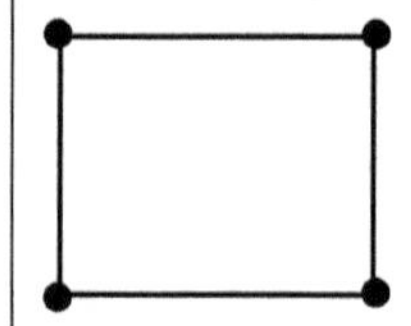

4.2 Paths and Circuits

Walks

Let G be a graph having at least one edge. In G, consider a finite, alternating sequence of vertices and edges of the form $v_i\, e_j\, v_{i+1}\, e_{j+1}\, v_{i+2}, \ldots, e_k\, v_m$. It starts and ends with the vertices and are such that each edge in the sequence is incident on the vertices preceding and following it in the sequence. Such a sequence is called a walk in G.

In a walk, a vertex or an edge or both could appear more than once. The number of edges present in a walk is called its 'length'.

Example:

Let us consider the graph shown below:

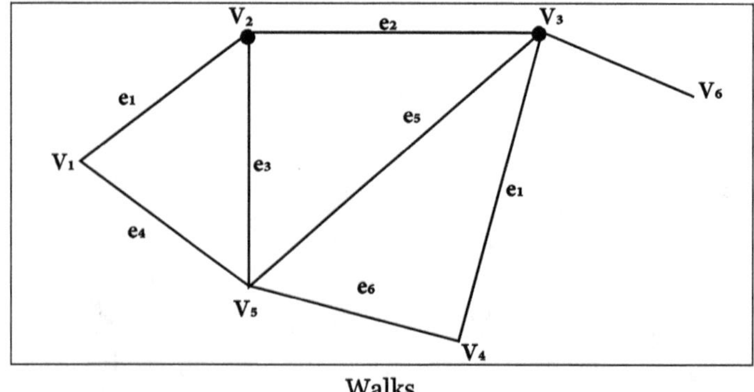

Walks

In this graph:

- The sequence $v_1\ e_1\ v_2 c_2\ v_3,\ e_8 v_6$ is a walk of length 3 (because this walk contains 3 edges; e_1, e_2, e_8). In this walk, no vertex and no edge is repeated.
- The sequence $v_1, e_4\ v_5 e_3\ v_2 c_2 v_3 e_5\ v_5 e_6 v_4$ is a walk of length 5. In this walk, the vertex v_5 is repeated, but no edge is repeated.
- The sequence $v_1 e_1 v_2 e_3 v_5 e_3 v_2 e_2 v_3$ is a walk of length 4. In this walk, the edge e_3 is repeated and the vertex v_2 is repeated.

A walk that begins and ends at the same vertex is called a closed walk. In other words, a closed walk is a walk in which the terminal vertices are coincident. A walk which is not closed is called an open walk. In other words, an open walk is a walk that begins and ends at two different vertices. For example, in the graph shown in figure, $v_1 e_1 v_2 c_3 v_5 e_4 v_1$ is a closed walk and $v_1 e_1 v_2 e_2 v_3 e_5 v_5$ is our open walk.

Paths

A walk of length k in a graph is a succession of k edges of the form uv, vw ,wx,... ,yz. This walk is denoted by uv, wx ... xz, and is referred to as a walk between u and z. A walk is closed if u=z.

Path

A sequence of distinct vertices such that two consecutive vertices are adjacent. Example: (a, d, c, b, e) is a path (a, b, e, d, c, b, e, d) is not a path; it is a walk. A trail in which no vertex appears more than once is called a path.

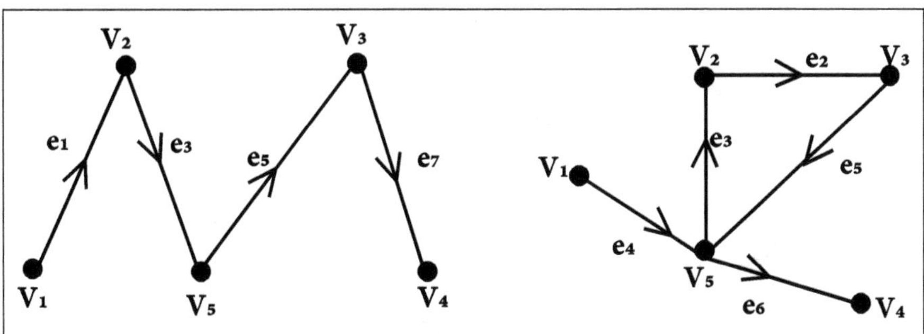

Path and Not a path.

The following facts are to be emphasized:

- A walk can be open or closed. In a walk, a vertex and/or an edge will appear over once.
- A trail is an open walk in which a vertex can appear more than once but an edge cannot appear more than once.

- A circuit is a closed walk in which a vertex can appear more than once but an edge cannot appear more than once.
- A path is an open walk in which neither a vertex nor an edge can appear more than once. Each path is a trail; but a trail need not be a path.
- A cycle is a closed walk in which neither a vertex nor an edge can appear more than once.

Circuits

In a walk, vertices and/or edges may appear more than once. If in an open walk no edge appears more than once, than the walk is called a trail. A closed walk in which no edge appears more than once is called a circuit.

For example:

In the figure, the open walk $v_1e_1v_2e_3v_5e_3v_2e_2v_3$ shown separately in figure is not a trail (because, in this walk, the edge e_3 is repeated) whereas the open walk $v_1e_4v_5e_3v_2v_2v_3e_5v_5e_6v_4$ shown separately in figure is trail.

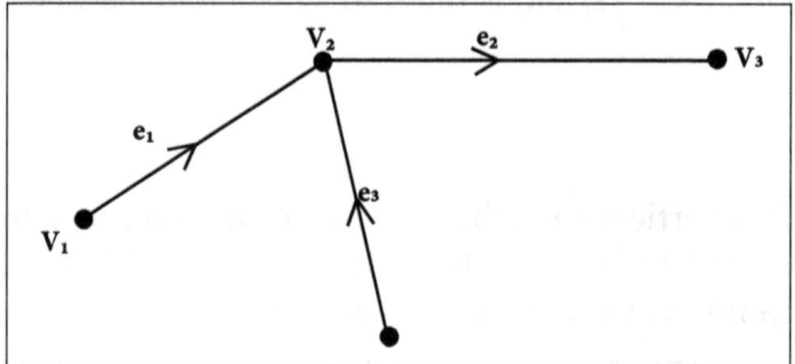

Not a trial.

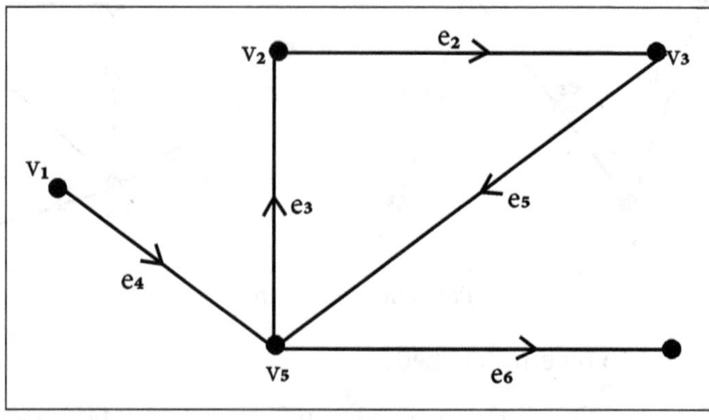

Trial

Also, in the same figure, the closed walk $v_1e_1v_2e_3v_5e_3v_2e_2v_3e_5v_5e_4v_1$ shown separately in figure is not a circuit (because e_3 is repeated) where as the closed walk $v_1e_1v_2e_3v_5e_5v_3e_7v_4e_6v_5e_4v_1$ shown separately in figure is a circuit.

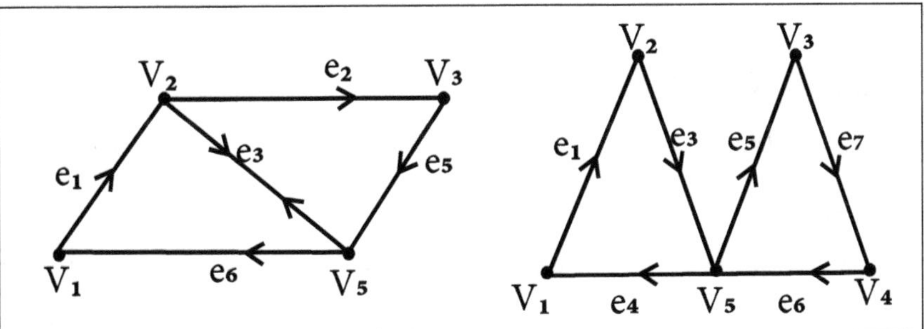

Not a circuit.

4.2.1 Graph Traversals

Depth - First Search Algorithm

- DFS starts visiting vertices of a graph at an arbitrary vertex by marking it as visited.
- It visits graph's vertices by always moving away from last visited vertex to an unvisited one, backtracks if no adjacent unvisited vertex is available.
- It is a recursive algorithm, it uses a stack.
- A vertex is pushed onto the stack when it is reached for the first time.
- A vertex is popped off the stack when it becomes a dead end, i.e., when there is no adjacent unvisited vertex.
- "Redraws" graph in tree-like fashion (with tree edges and back edges for undirected graph).

The Strategy

- Starting vertex may be determined by the problem or chosen arbitrarily.
- Visits graph's vertices by always moving away from last visited vertex to unvisited one, backtracks if no adjacent unvisited vertex is available.
- Uses a stack representation:
    - A vertex is pushed onto stack when it is reached for the first time.
    - A vertex is popped off from the stack when it becomes a dead-end, i.e. there are no adjacent unvisited vertices.
    - Redraws the graph in tree-like fashion (with tree edges and back edges for undirected graph). The theme of the DFS is to explore if possible otherwise backtrack.

Algorithm

ALGORITHM DFS (G)

```
// Implements a depth-first search traversal of a given graph.

// Input: Graph G = (V, E).

// Output: Graph G with its vertices marked with consecutive integers.

// in the order they have been first encountered by the DFS traversal
mark each vertex in V with 0 as a mark of being "unvisited"

Count ← 0

For each vertex v in V do

If v is marked with 0

dfs (v)

dfs (v)

//visits recursively all the unvisited vertices connected to vertex v
and

//assigns them the numbers in the order they are encountered

//via global variable count.

Count ← count + 1; mark v with count

For each vertex w in V adjacent to v do

If w is marked with 0

dfs (w)
```

Example:

Given graph:

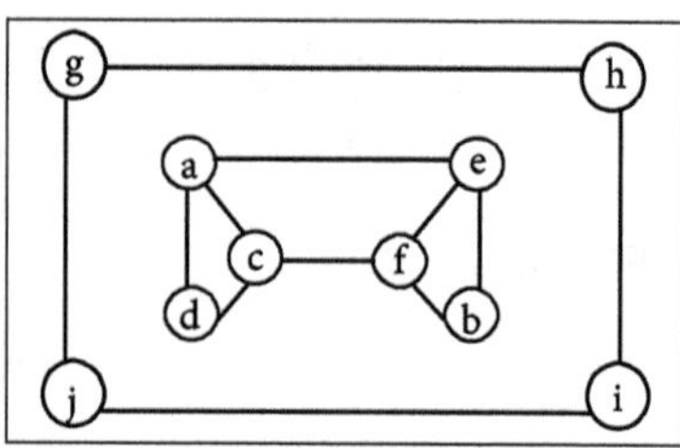

Traversal using DFS

- Let the start vertex is a and the traversal is carried out, by selecting the adjacent vertex in alphabetical order of preference (i.e.. if more than one vertex is adjacent to start vertex).

- Initially all the vertices are assigned with 0, while traversing each vertex is numbered indicating the order in which a vertex was visited and pushed onto the stack.
- When the path through the vertex becomes a dead-end, it is popped off from the stack and the order is numbered:

Vertex	Pushed onto the stack	Popped off from the stack
a	1	6
c	2	5
d	3	1
f	4	4
b	5	3
e	6	2
g	7	10
h	8	9
i	9	8
j	10	7

  - From node d, a edge to node a (visited vertex). Therefore it is popped off from the stack and returned to c.
  - From c the nodes f, b, e are traversed. From e, path exists only for visited vertices (b, a). Therefore e is popped off from the stack and returned to b.
  - From b on wards the vertices f, c, a are popped off from the stack.
  - The nodes g, h, i, j are traversed with gas root and popped off, generates a depth first spanning forest (DFS forest).
- DFS Forest: Consists of two DFS trees, where tree edges are shown with solid lines and back edges are represented with dashed lines.

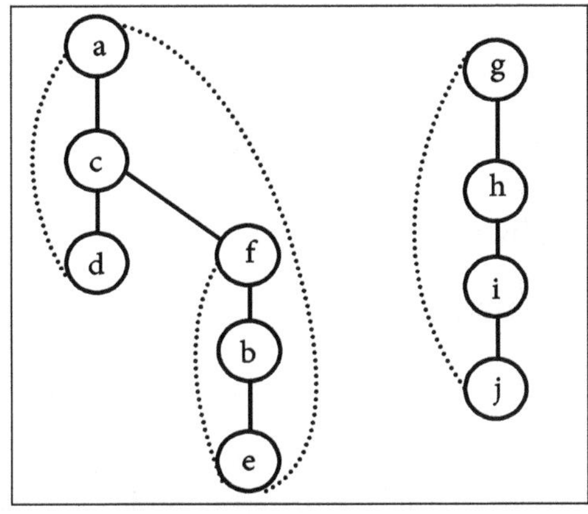

DFS trees of given graph.

Applications of DFS:

- The two orderings are advantageous for various applications like topological sorting, etc.
- To check connectivity of a graph.
- To check if a graph is acyclic.
- To find articulation point in a graph.

Efficiency:

Depends on the graph representation:

- Adjacency matrix : $\Theta(n^2)$
- Adjacency list: $\Theta(n + e)$

Breadth-First Search (BFS)

- BFS starts visiting vertices of a graph at an arbitrary vertex by marking it as visited.
- It visits graph's vertices by across to all the neighbors of the last visited vertex.
- Instead of a stack, BFS uses a queue.
- Similar to level-by-level tree traversal.
- "Redraws" graph in tree-like fashion (with tree edges and cross edges for undirected graph).

The Strategy

- Starting vertex may be determined by the problem or chosen arbitrarily.
- Visits graph vertices by moving across to all the neighbors of last visited vertex.
- Uses a queue representation:
    - A vertex is inserted into queue when it is reached for the first time, marked as visited.
    - A vertex is removed from the queue, when it identifies all unvisited vertices

that are adjacent to the vertex, marked as visited and added into the queue.

- The order of visiting nodes level-by level is BFS traversal.
- Redraws graph in tree-like fashion (with tree edges and cross edges for undirected graph).

Algorithm BFS (G)

```
//Implements a breadth-first search traversal of a given graph

//Input: Graph G= (V, E)

//Output: Graph G with its vertices marked with consecutive integers

//in the order they have been visited by the BFS traversal marked each
vertex in V with 0 as a mark of being "unvisited"

Count ← 0

For each vertex v in V do

If v is marked with 0

bfs (v)

bfs (v)

//visits all the unvisited vertices connected to vertex v

//and assigns them the numbers in the order they are visited

//via global variable count

count ← count + 1; mark v with count and initialize a queue with v

While the queue is not empty do

For each vertex w in V adjacent to the front's vertex v

do

If w is marked with 0

count ← count + 1; mark w with count

add w to the queue

remove vertex v from the front of the queue
```

Example:

Given graph:

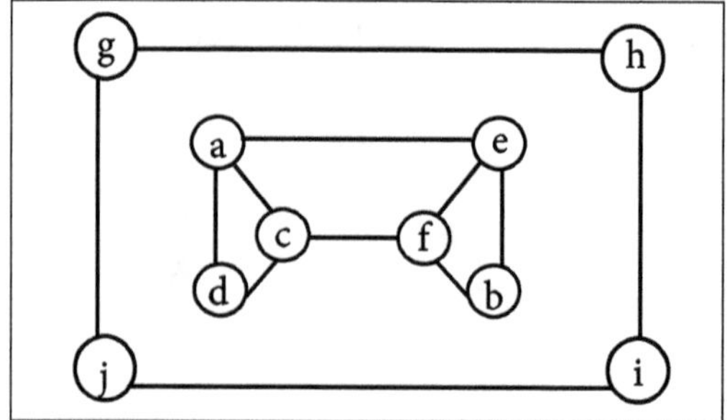

BFS trees:

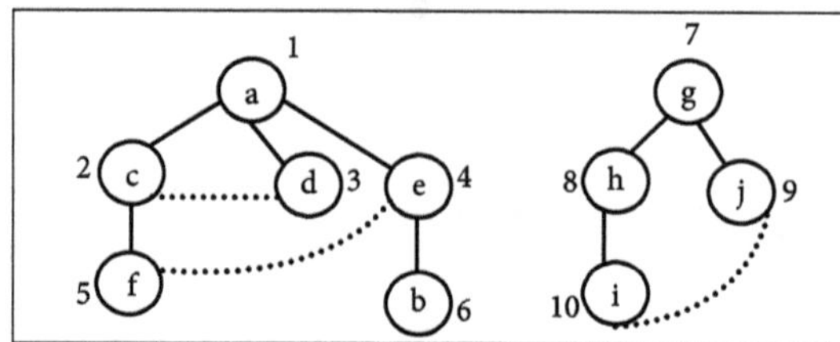

BFS trees of given graph.

Applications of BFS:

- To check connectivity of a graph (number of times queue becomes empty tells the number of components in the graph).
- To check if a graph is acyclic (no cross edges indicates no cycle).
- To find minimum edge path in a graph.

Efficiency

Depends on the graph representation:

- Array : $\Theta(n^2)$
- List: $\Theta(n + e)$

Depth-First Search (DFS) vs Breadth-First Search (BFS)

	DFS	BFS
Data structure	stack	queue
No. of vertex orderings	2 orderings - push and pop of stack.	1 ordering - Insertion and deletion from the queue is same.

Edge types (undirected graphs)	tree and back edges	tree and cross edges
Applications	Connectivity, acyclicity, articulation points.	connectivity, acyclicity, minimum-edge paths.
Efficiency for adjacent matrix	$\Theta(\|V^2\|)$	$\Theta(\|V^2\|)$
Efficiency for adjacent linked lists	$\Theta(\|V\|+\|E\|)$	$\Theta(\|V\|+\|E\|)$

4.2.2 Shortest Path in Weighted Graphs

The shortest path problem is the problem of finding a path between two vertices in a graph such that the sum of the weights of its constituent edges is minimized.

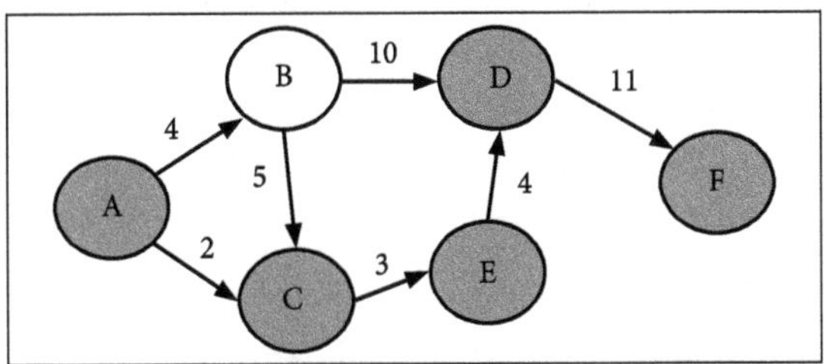

Types of Graph:

1. Weighted: A graph is a weighted graph if a number (weight) is assigned to each edge.

Example: Such weights might represent costs, lengths or capacities, etc.

2. Un-weighted: A graph is a un-weighted graph if a number (weight) is not assigned to each edge.

3. Directed: A directed graph is one in which edges have orientation.

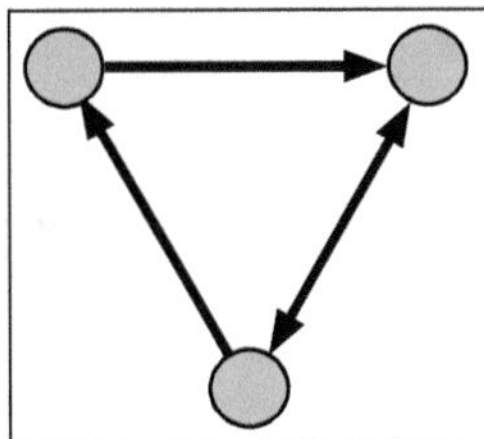

4. Un-directed: An undirected graph is one in which edges have no orientation.

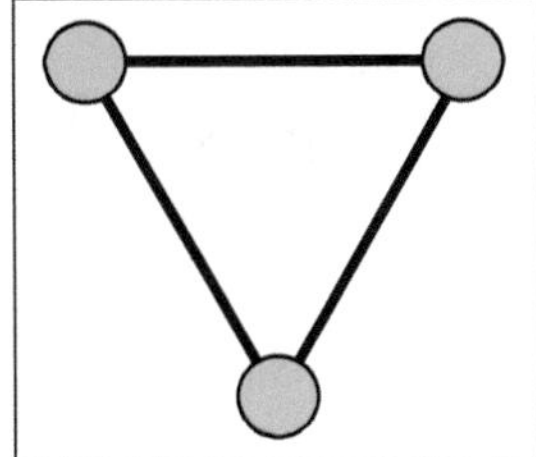

Dijkstra's Algorithm:

- Works when all of the weights are positive.
- Provides the shortest paths from a source to all other vertices in the graph.
- Can be terminated early once the shortest path to t is found if desired.

4.2.3 Euler Ian Paths and Circuits

Consider a connected graph G. If a circuit in G contains all the edges of G then that circuit is called an Euler circuit (or Eulerian line or Euler tour) in G. If there is a trail in G that contains all the edges of G, then that trail is called an Euler trail.

In a trail and a circuit, no edge can appear more than once, but a vertex can appear more than once. This property, which is carried to Euler trails and Euler Circuits also. Since the Euler circuits and Euler trails include all edge and then automatically should include all vertices as well.

A connected graph which has an Euler circuit is called a Semi-Euler graph (or a Semi Eulerian graph). For example, in the graph shown in figure closed walk, $Pe_1Qe_2Re_3Pe_4Se_5Re_6Te_7P$ is an Euler circuit. Therefore, this graph is an Euler graph.

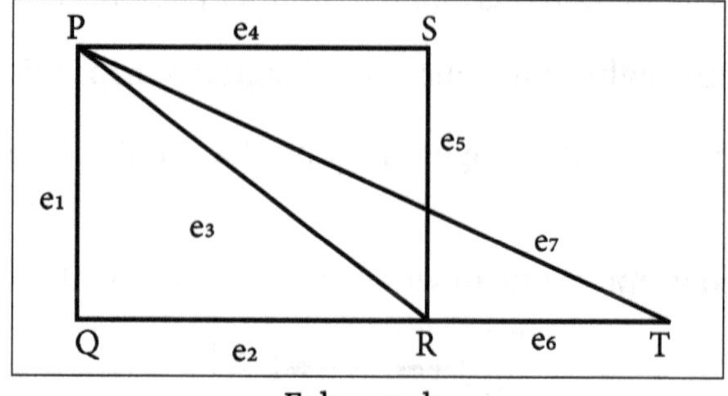

Euler graph.

Let us consider the graph shown in the figure. In this graph, every sequence of edges which starts and ends with the same vertex and includes all the edges will contain at least one repeated edge. Thus, the graph has no Euler circuits. Hence, this graph is not an Euler graph.

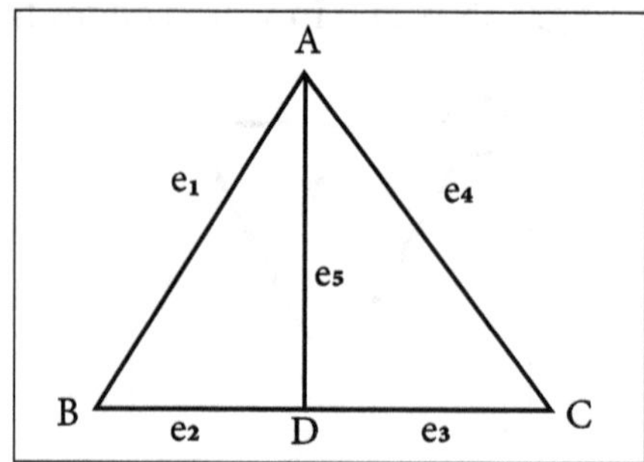

It might be seen that the trail $Ae_1Be_2De_3Ce_4Ae_5D$ in the graph is an Euler trail. This graph is therefore a Semi-Euler Graph.

Theorem

A given connected graph G is an Euler graph if and only if all vertices of G are of even degree.

Proof:

Suppose that G is an Euler graph. It therefore contains an Euler line. In tracing this walk, we observe that every time the walk meets a vertex v, it goes through two "new" edges incident on v with one "entered" v and with the other "exited".

This is true not only for all the intermediate vertices of the walk, but also for the terminal vertex, because we "exited" and "entered" the same vertex at the beginning and end of the walk, respectively. Thus if G is an Euler graph, the degree of every vertex is even.

To prove the sufficiency of the condition, assume that all the vertices of G are of even degree. Now, we construct a walk starting at an arbitrary vertex v and going through the edges of G such that no edge is traced more than once. We continue tracing as far as possible.

Since every vertex is of even degree, we can exit from every vertex we enter. The tracing cannot stop at any vertex, but at v. And since v is also of even degree, we shall eventually reach v when the tracing comes to an end. If this closed walk h, we just traced includes all the edges of G, G is an Euler graph.

If not, we remove all the edges in h from G, and obtain a sub graph h' of G formed by the remaining edges. Since both the G and h have all their vertices of even degree, the degrees of the vertices of h' are also even. Moreover, h' must touch h at least at one vertex a, because G is connected. Starting from a, we can again construct a new walk in graph h'.

Since all the vertices of h' are of even degree, this walk in h' must terminate at vertex a. But this walk in h' can be combined with h to form a new walk, which starts and ends at vertex v and has more edges than h. This process can be repeated until we obtain a closed walk that traverses all the edges of G. Thus, G is an Euler graph.

Unicursal Graph

An open walk that includes all the edges of a graph without retracing any edge is called unicursal line or an open Euler line. A (connected) graph that has a unicursal line will be called a unicursal graph.

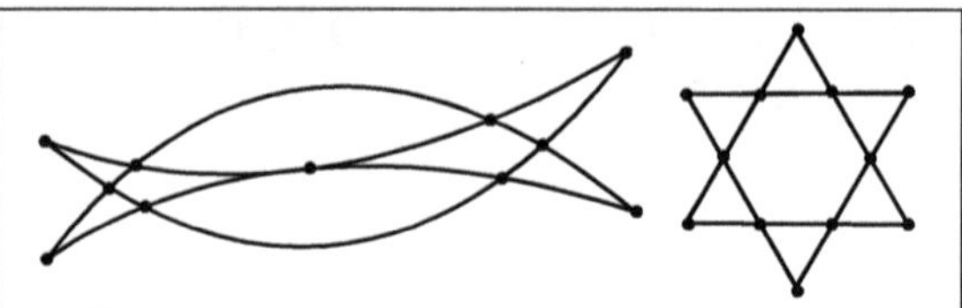

Euler Graphs

- Mohammed's scimitars.
- Star of david.

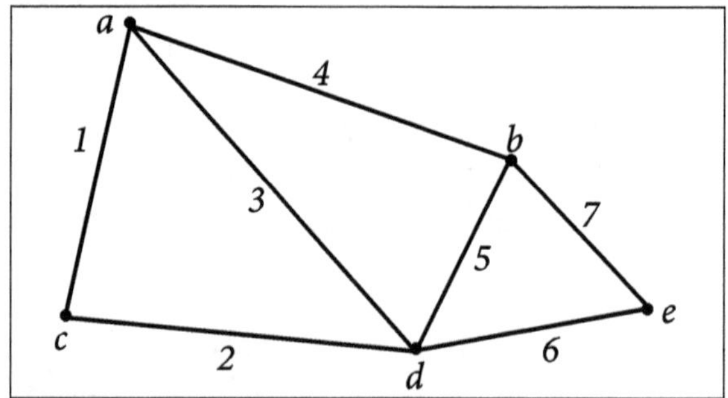

Theorem

In a connected graph G with exactly 2k odd vertices, there exist k edge-disjoint sub graphs such that they together contain all edges of G and that each is a unicursal graph.

Proof:

Let the odd vertices of the given graph G is named as v_1, v_2, ..., v_k; w_1, w_2, ...w_k in any arbitrary order. Add k edges to G between the vertex pairs (v_1, w_1), (v_2, w_2), ..., v_k, w_k) to form a new graph G'. Since every vertex of G' is of even degree, G' consists of an Euler line p.

Now if we remove the k edges from p we just added, p will be split into k walks, each of which is a unicursal line. The first removal will leave a single unicursal line. The second removal will split that into two unicursal lines and each successive removal will split the unicursal line into two unicursal lines, until there are k of them. Thus, the theorem is proved.

Fleury's Algorithm

To find an Euler path or an Euler circuit:

- Make sure the graph has either 0 or 2 odd vertices.
- If there are 0 odd vertices, start anywhere. If there are 2 odd vertices, start at one of them.

- Follow edges one at a time. If we have a choice between a bridge and a non-bridge, always choose the non-bridge.
- Stop when we run out of edges. This is called Fleury's algorithm.

Problems

1. Let us show that the following graph contains an Euler Circuits.

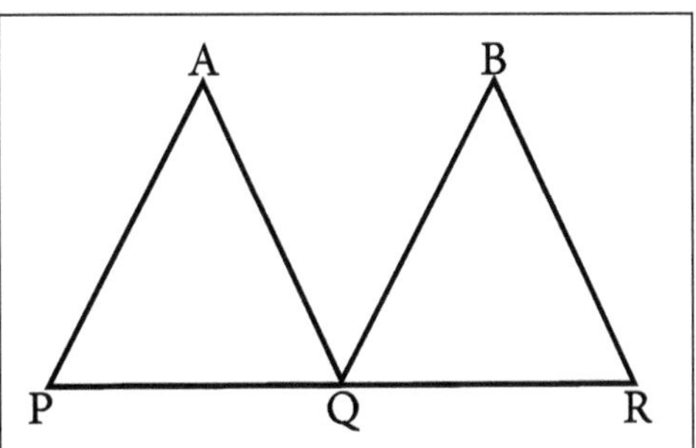

Solution:

Given:

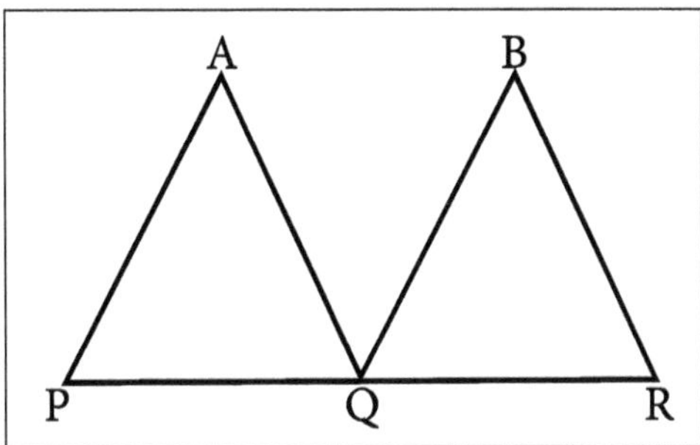

Here if the trail starts from P each and every node is visited at least once and ends with starting edge P. The graph contains an Euler Circuit PAQBRQP.

2. Let us determine an Euler circuit in the graph below.

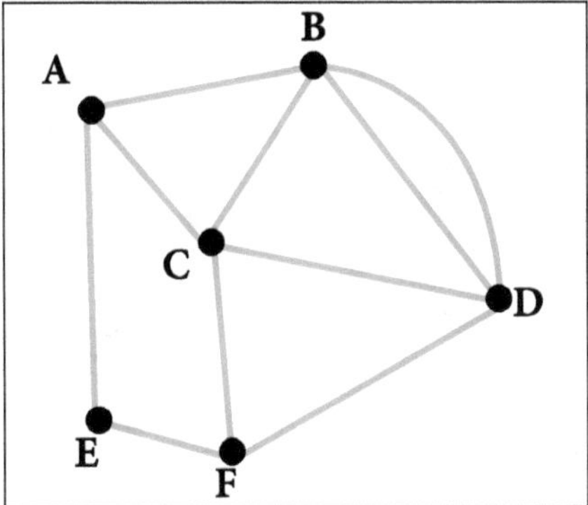

Solution:

There are two odd vertices, A and F. Let us start at F.

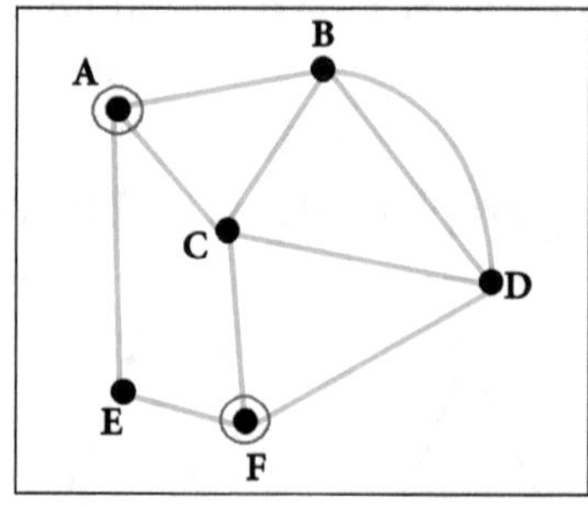

Start walking at F. When we use an edge, delete it.

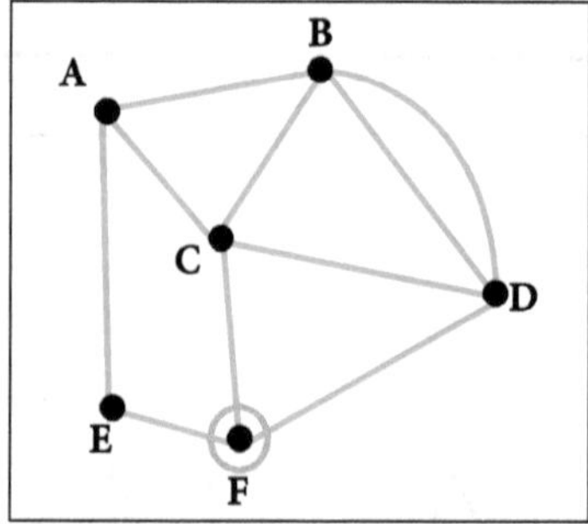

Path so far: FE.

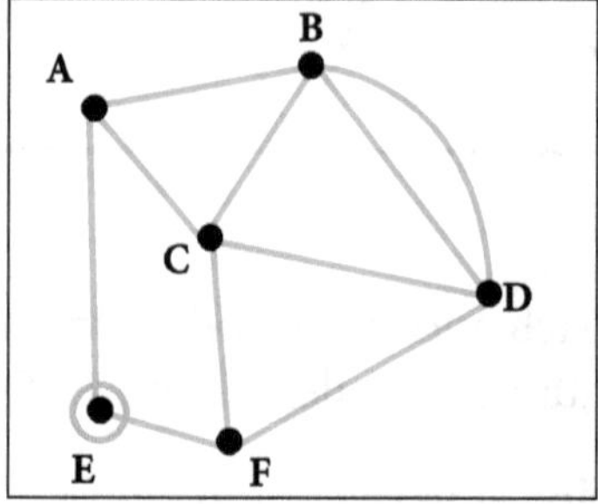

Path so far: FEA.

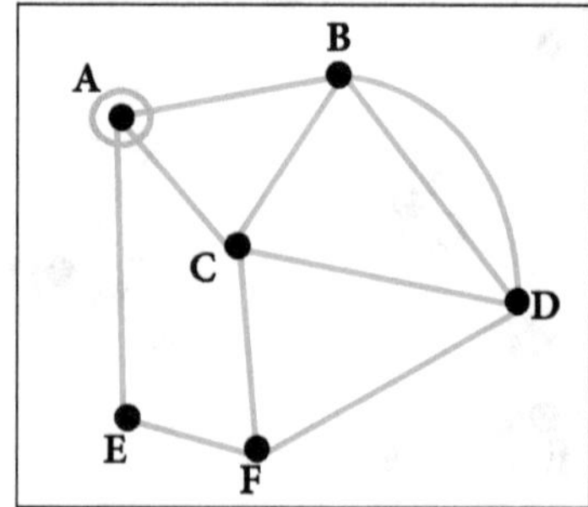

Path so far: FEAC.

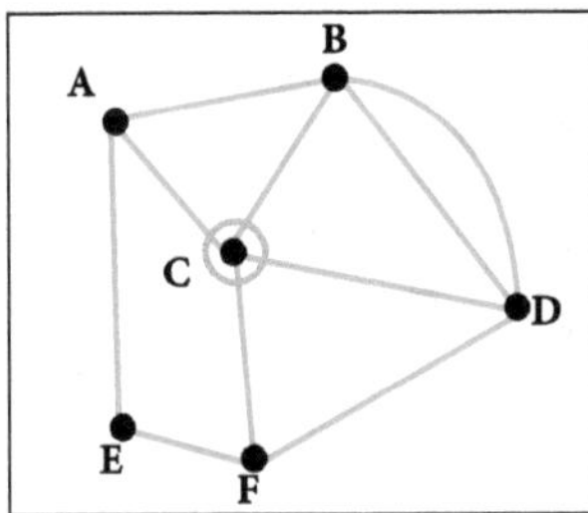

Path so far: FEACB.

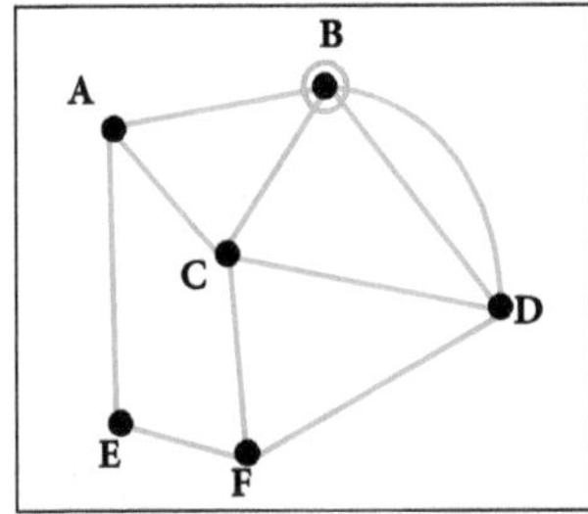

Until this point, the choices did not matter.

But now, crossing the edge BA would be a mistake, because we would be stuck there. The reason is that BA is a bridge. We don't want to cross a bridge unless it is the only edge available.

Path so far: FEACB.

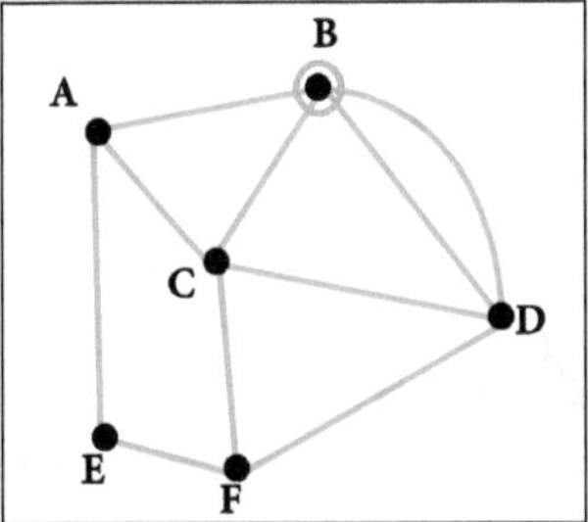

Path so far: FEACBD.

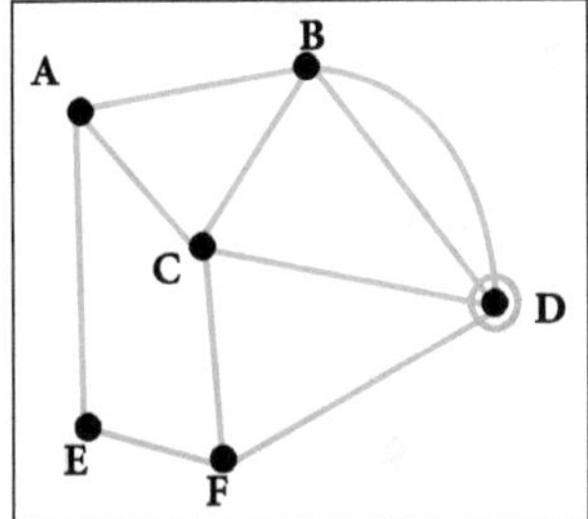

Path so far: FEACBD. Don't cross the bridge.

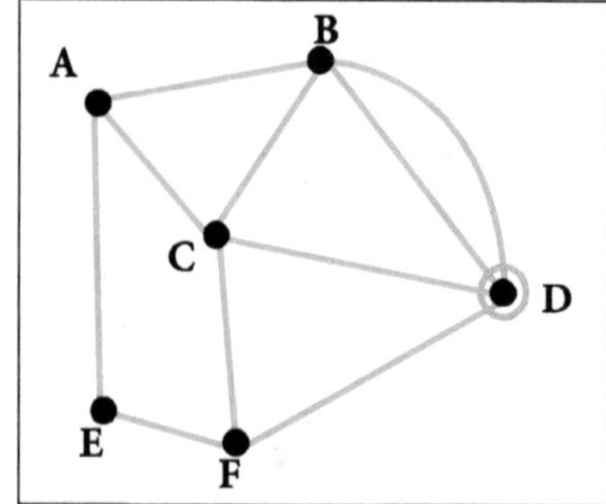

Path so far: FEACBDC.

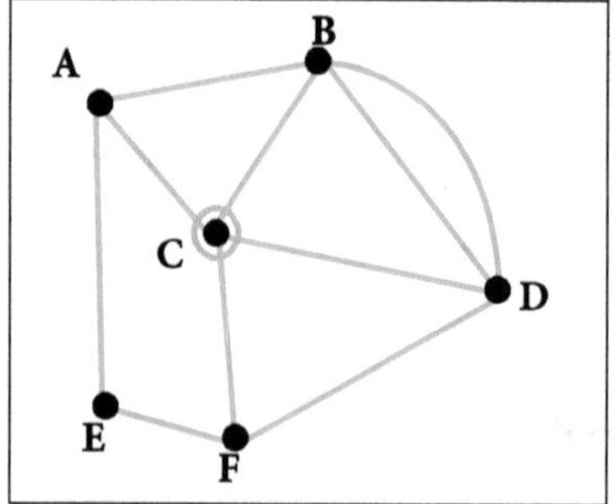

Path so far: FEACBDC. Now we have to cross the bridge CF.

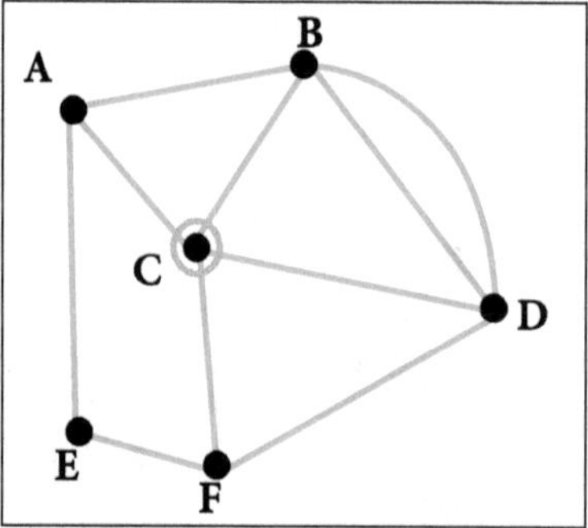

Path so far: FEACBDCF.

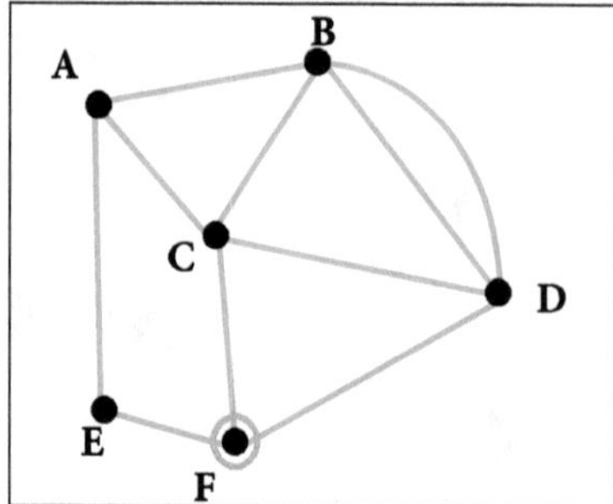

Path so far: FEACBDCFD.

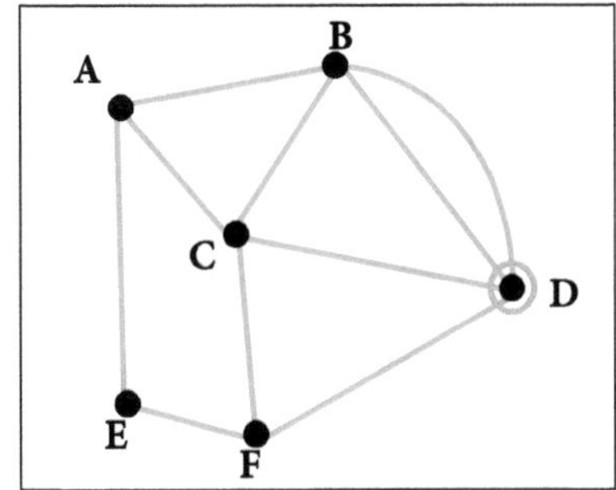

Path so far: FEACBDCFDB.

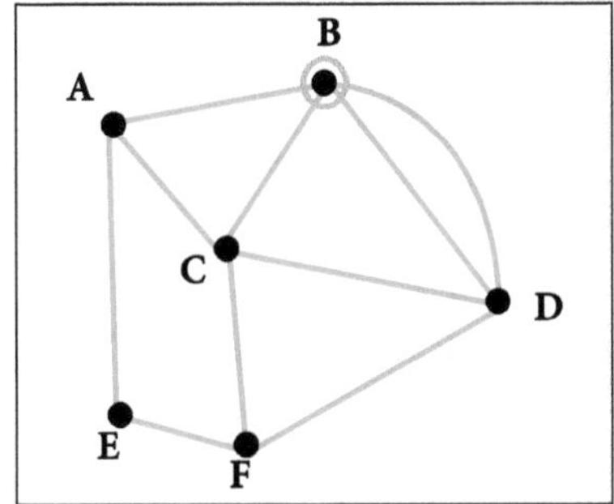

Euler Path: FEACBDCFDBA.

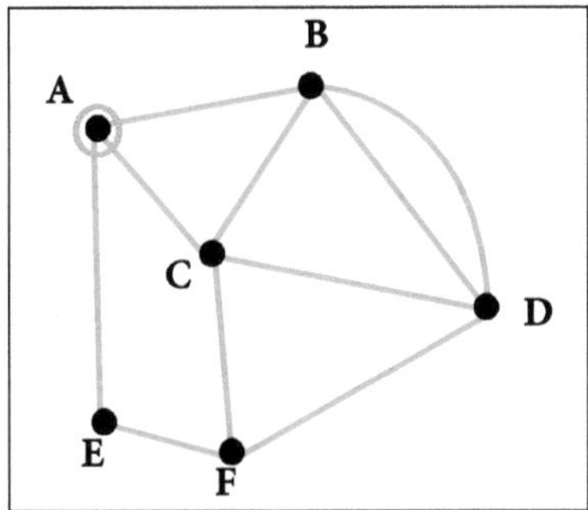

Euler Path: FEACBDCFDBA.

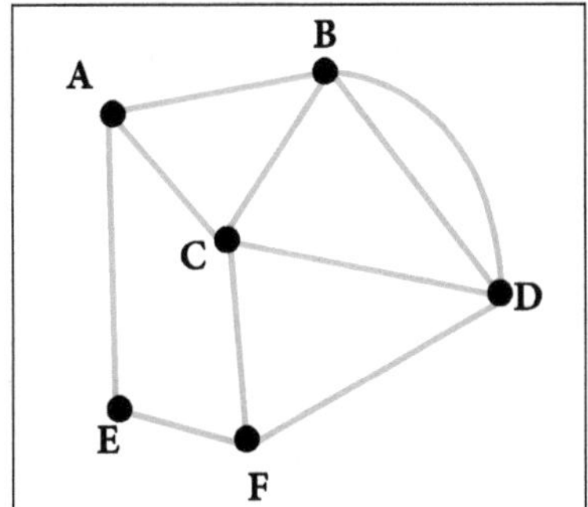

4.3 Hamiltonian Paths and Circuits

If G is not Eulerian, the poor postman has to walk at least one street twice. This happens, e.g., if one of the streets is a dead end and in general, if there is a street corner of an odd number of streets. We can solve this case by reducing it to the Eulerian case as follows:

An edge e = uv will be duplicated, if it is added to G parallel to an existing edge e′= uv with the same weight, $\alpha\ (e') = \alpha\ (e)$.

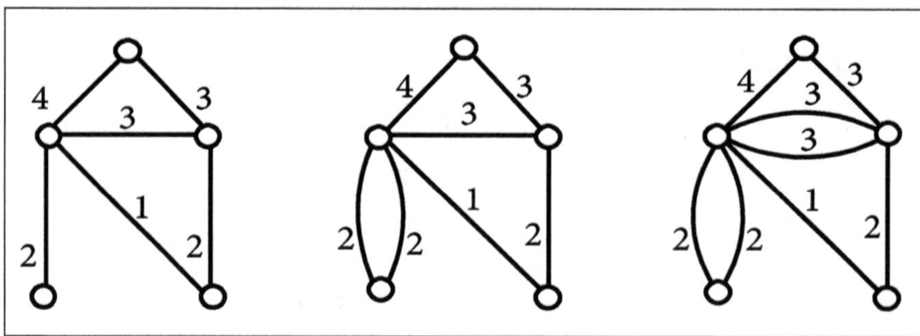

Above we have duplicated two edges. The rightmost multi-graph is Eulerian. There is a good algorithm by EDMONDS AND JOHNSON (1973) for the construction of an optimal Eulerian super graph by duplication. Unfortunately, this algorithm is somewhat complicated.

Hamiltonian Graphs

In the connector problem, we reduce the cost of a spanning graph to its minimum. There are different problems, where the cost is measured by an active user of the graph. For instance, in the traveling salesman problem, a person is supposed to visit each town in his district and this should do in such a way that it saves time and money. Obviously, the travel should be planned so as to visit each town once, so that the overall flight time is as short as possible.

In terms of graphs, he is looking for a minimum weighted Hamilton cycle of a graph, the vertices of which are the towns and the weights on the edges are the flight times. Unlike for the shortest path and for connector problems, no efficient reliable algorithm is known for the traveling salesman problem. Indeed, it is widely believed that no practical algorithm exists for this problem.

Hamilton Cycles

A path P of a graph G is the Hamilton path, if P visits every vertex of G once. Similarly, a cycle C is a Hamilton cycle, if it visits every vertex once. A graph is Hamiltonian, if it has the Hamilton cycle.

Note that if C: $u_1 \rightarrow u_2 \rightarrow \ldots \rightarrow u_n$ is a Hamilton cycle, so is $u_i \rightarrow \ldots u_n \rightarrow u_1 \rightarrow \ldots u_{i-1}$ for each $i \in [1, n]$, thus we can choose where to start the cycle.

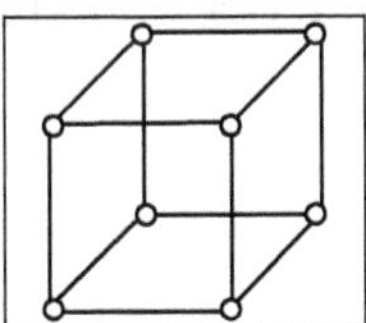

If G is Hamiltonian, then for every non-empty subset $S \subseteq V_G$, $c(G-S) \leq |S|$.

Proof:

Let $\varnothing \neq S \subseteq V_G$, $u \in S$ and let C: u, u be a Hamilton cycle of G.

Let us assume G−S has k connected components, G_i, $i \in [1, k]$. The case k = 1 is trivial and suppose that k > 1.

Let u_i be the last vertex of C that belongs to G_i and let v_i be the vertex that follows u_i in C. Now $v_i \in S$ for each i by the choice of u_i and $v_j \neq v_t$ for all $j \neq t$, because C is a cycle and $u_i v_i \in G$ for all i. Thus $|S| \geq k$ as required.

Theorem

In a complete graph with n vertices, there are (n-1)/2 edge-disjoint Hamiltonian circuits, if n is an odd number ≥3.

Proof:

A complete graph G of the n vertices has n (n -1)/2 edges and the Hamiltonian circuit in G consists of n edges. Therefore, the number of edge-disjoint Hamiltonian circuits in G cannot exceed (n-1)/2. There are (n -1)/2 edge-disjoint Hamiltonian circuits, when n is odd, can be shown as follows:

The sub graph in the below figure is a Hamiltonian circuit.

Keeping the vertices fixed on the circle, rotate the polygonal pattern clockwise by 360/(n - 1), 2x360/ (n - 1), 3x360/ (n - 1) .... (n- 3)/2x360/(n-1) degrees.

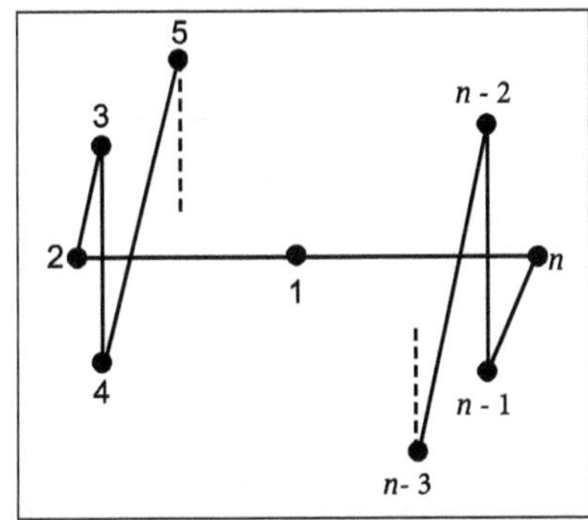

Observe that each rotation produces a Hamiltonian circuit that has no edge in common with any of the previous ones. Thus, we have (n - 3)/2 new Hamiltonian circuits, in which all the edge disjoint from the one in figure and also the edge disjoint among themselves. Hence, the theorem is proved.

Problems

1. Let us solve, $X = \{1, 2, ..., 7\}$ and $R = \{(x, y) \mid x - y\}$ is divisible by 3. Also show that R is an equivalence relation and also let's draw the graph of R.

Solution:

Given:

$$X = \{1, 2, ..., 7\} \text{ and } R = \{(x, y) \mid x - y\}$$

$$\text{Let } R = \{(1,1), (1,4), (1,7), (2,2), (2,5), (3,3), (3,6), (4,4), (4,1), (4,7),$$

$$(5,2), (5,5), (6,3), (6,6), (7,1), (7,4), (7,7)\}.$$

- x - x = 0 is divisible by 3 for all x ∈ X

∴ R is reflexive.

- When x - y is divisible by 3 then y - x is also divisible by 3. Hence, R is symmetric.
- When x - y and y - z are divisible by 3. Then:

(x - y) + (y - z) = x - z is also divisible by 3.

For x, y, z ∈ X.

∴ R is transitive.

Hence, R is reflexive, symmetric and transitive.

∴ R is an equivalence relation.

Graph of R

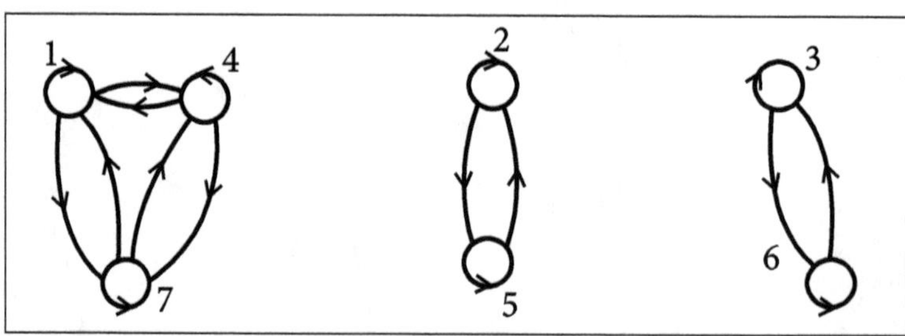

2. Let us G be a simple in directed graph with n vertices. Let u and v be two non-adjacent vertices in G such that deg (u) + deg (v). ≥ n in G. Let us show that G is Hamiltonian if and only if G + uv is Hamiltonian.

Solution:

Given:

The graph G is a spanning subgroup of G + uv. So if G is hamiltonian then G + uv is also hamiltonian.

Conversely

Assume that G + uv is hamiltonian. To prove that G is hamiltonian.

If G + uv has a Hamilton cycle, C not containing the edge uv, then C itself is a Hamilton cycle in G.

So assume that every Hamilton cycle C of G + uv contains the edge uv. If C is a Hamilton cycle of G + uv containing the edge uv. Then C - uv is a uv Hamilton path in G.

So we have a Hamilton path:

$$u = v_1, v_2, \ldots v_n = v \text{ in G.}$$

Let $S = \{v_i / uv_i + 1 \in E(G) : 1 \le I \le n-1\}$ and $T = \{v_i / v_i v \in E(G)\}$

Clearly $v = v_n \notin SUT$. So $|SUT| \le n-1$

$$|S| + |T| = \deg(u) + \deg(v) \ge n.$$

So, from $|SUT| + |S \cap T| = |S| + |T|$, it follows that $|S \cap T| \ge 1$ and $S \cap T \ne \phi$.

Let $V_i \in S \cap T$. Then $uv_i + 1$and vv_i-1 are edges in G and uv + 1 v_i + 2 ... vv_iv_i - 1 ... v_2u is a Hamilton cycle.

Hence, G is Hamiltonian.

3. Let us draw the graph with 5 vertices A, B, C, D and E such that the deg (A) = 3, B is an odd vertex, deg (C) = 2 and D and E are adjacent.

Solution:

Given:

$$\deg(A) = 3; \deg(C) = 2$$

D & E are adjacent; B odd vertex.

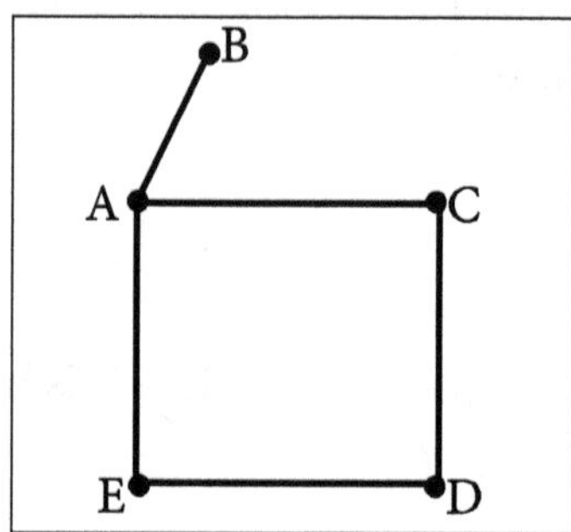

4.3.1 Traveling Sales Person's Problem

We saw a constant factor approximation to Minimum Bin Packing. However, it does not have a PTAS unless P = NP. This shows the direction PTAS = APX ⇒ P = NP. One example of a problem that cannot be in APX unless P = NP is that of well-known Minimum Traveling Salesman problem.

Minimum Traveling Salesman

Instance: A set $C = \{c_1, \ldots\ldots, c_n\}$ of cities and a distance function d:

$$C \times C \rightarrow N$$

Solution:

A path through the cities, i.e. A permutation $\pi : \{1, \ldots\ldots, n\} \rightarrow \{1, \ldots\ldots, n\}$.

Measure: The cost of visiting cities with respect to the path, i.e.

$$\sum_{i=1}^{n-1} d\left(c_{\pi(i)}, c_{\pi(i+1)}\right)$$

It is important to note that when the distances in the problem instances always obey a Euclidean metric, Minimum Traveling Salesperson has PTAS. Thus, we may say that it is the generality of possible distances in the above problem that makes it difficult to approximate. This is often the case with approximability: a small restriction on an in approximable problem can suddenly turn it into a highly approximable one.

Analysis of Traveling Salesperson

- Let N be the number of g (i, S) s that have to be computed before the equation may be used to compute g (1, V – {1}).
- For each value of |S|, there are n – 1 choices of i.
- The number of distinct sets S of size k not including 1 and i is $\binom{n-2}{k}$.

Hence, $N=\sum_{k=0}^{n-2}(n-k-1)\binom{n-1}{k}=(n-1)2^{n-2}$.

- An algorithm that finds an optimal tour using the equations will require $\Theta(n^2 2^n)$ time as the computation of g(i, S) with |S| = k requires k – 1 comparisons when solving the equation.
- Better than enumerating all n! Different tours to find the best one.
- The most serious drawback of the dynamic programming solution is the space needed as O ($n2^n$). This can be too large even for modest values of n.

4.4 Planar Graphs

If a graph is said to be planar if it can be drawn in the plane in such a way that no 2 edges intersect each other. Drawing a graph in the plane without edge crossing is called embedding the graph in the plane.

Given a planar representation of a graph G, a face (also called a region) is a maximal section of the plane in which any 2 points can be joint by a curve that does not intersect any part of G.

When we trace around the boundary of a face in G, we encounter a sequence of vertices and edges, finally returning to our final position. Let v_1, e_1, v_2, e_2, . . . , v_d, e_d, v_1 be the sequence acquired by tracing around a face, then d is the degree of the face.

Some edges may be encountered two times because both the sides of them are on same face. A tree is an extreme example of this, each edge is being encountered twice.

Planar Graphs Representation

It has been denoted that a graph may be referred by more than one geometrical drawing. In some drawing representing graphs the edges intersect at points which are not vertices of the graph and in some others the edges meet only at the vertices.

A graph that can be represented by at least one plane drawing in which the edges meet only at the vertices is known as a 'planar graph'. On the other hand, a graph that cannot be represented by a plane drawing in which the edges meet only at the vertices is known as a non-planar graph. In other words, a non-planar graph is a graph whose every possible plane drawing has at least 2 edges which intersect each other at points other than vertices.

Definition

A graph G is a planar graph, if it has a plane figure P(G), known as the plane embedding of G, where the lines corresponding to the edges does not intersect each other except at their ends. The complete bipartite graph $K_{2,4}$ is called the planar graph.

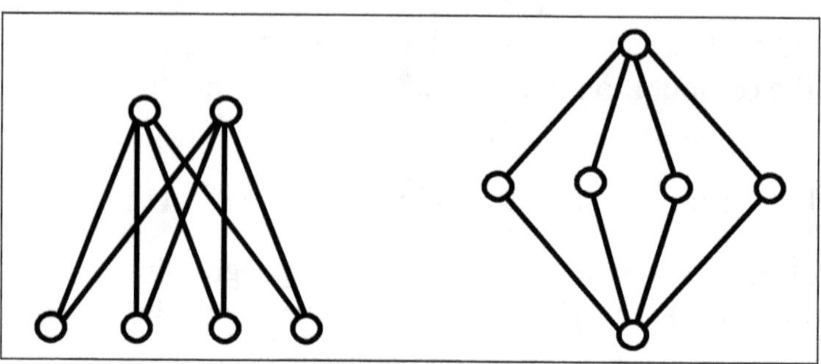

Bipartite graph.

An edge e = u v ∈ G is subdivided, when it is replaced by a path u –→ x –→ v of length two by introducing a new vertex x.

A subdivision H of a graph G is obtained from G by a sequence of subdivisions.

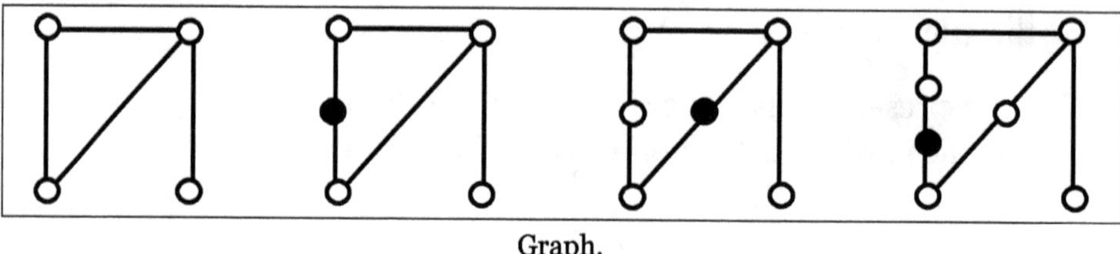

Graph.

Different Representation of a Planer Graph

A planar graph is a graph that can be drawn so that the edges only touch each other where they meet at vertices.

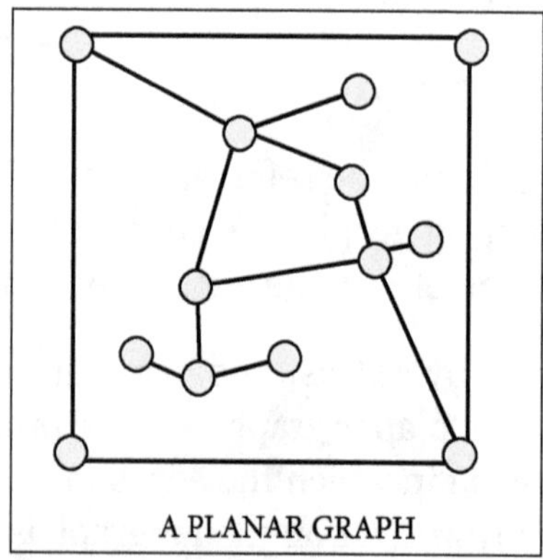

A PLANAR GRAPH

Planar graph.

We can generally redraw a planar graph so that some of the edges cross. Even so, it is still a planar graph. When it is drawn so that the edges cross, the drawing is called a non-planar representation of a planar graph.

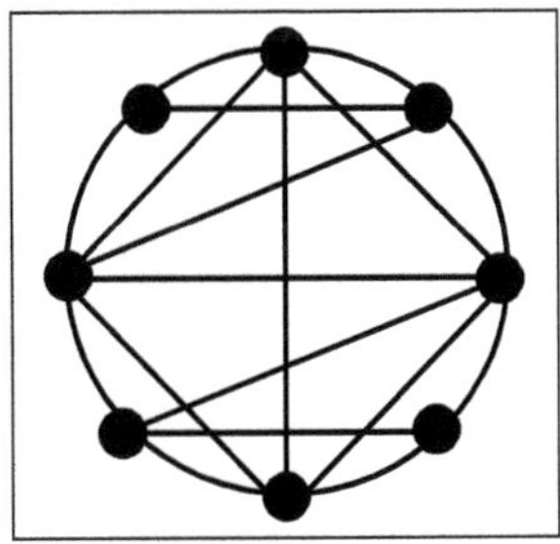

Non planar graph.

The geometry of the plane will be treated intuitively. A planar graph will be a graph that can be drawn in the plane so that no 2 edges intersect with each other. Such graphs are used in the design of electrical circuits, where one tries to avoid crossing the wires or laser beams.

Planar graphs can be used in some parts of mathematics, such as group theory and topology. There are fast algorithms for testing whether a graph is planar or not. However, the algorithms are all rather complex to implement. Most of them are based on an algorithm.

Different Representation of a Planar Graph

A planar graph is a graph that can be drawn so that the edges only touch each other where they meet at vertices.

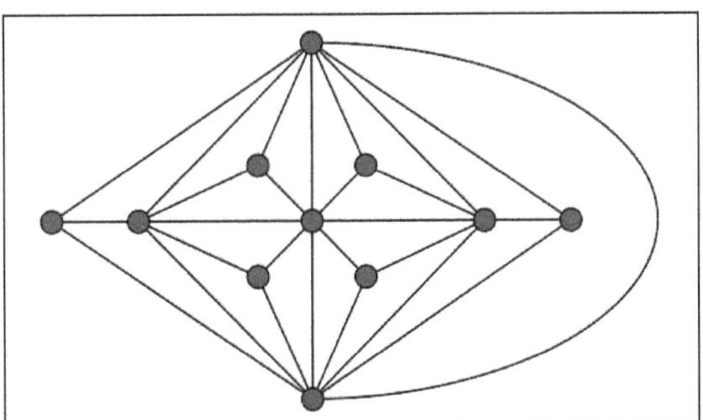

Planar graph.

We can usually redraw a planar graph so that some of the edges cross. Even so, it is still a planar graph. When it is drawn so that the edges cross, the drawing is called a non-planar representation of a planar graph.

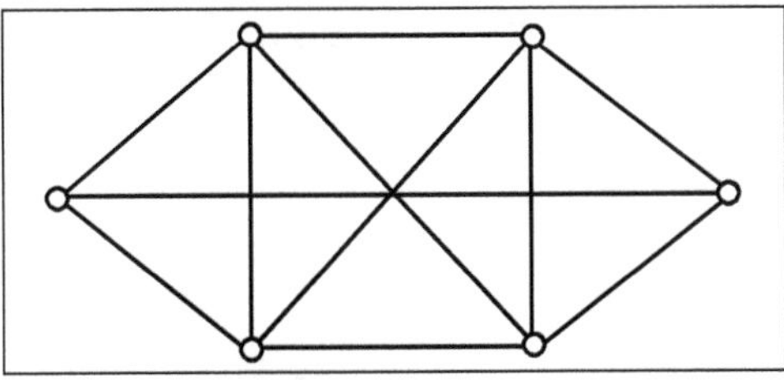

Non planar graph.

Embedding Graph

A drawing of a geometric representation of a graph on any surface such that no edges intersect is called embedding.

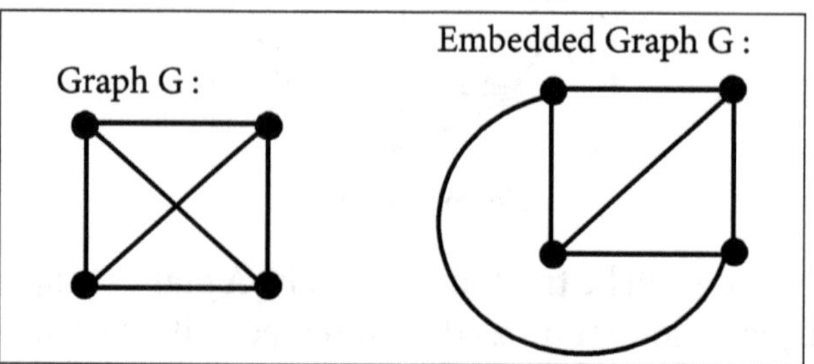

Regions in Graph

In any planar graph, drawn with no intersections, the edges divide the planes into different regions (faces, windows or meshes). The regions enclosed by the planar graph are known as interior faces of the graph. The region surrounding the planner graph is referred the exterior face of the graph.

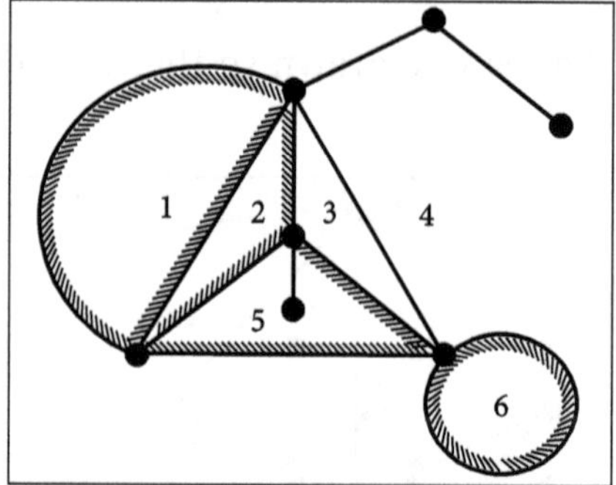

The graph has 6 regions.

Graph Embedding on Sphere

To eliminate the distinction between the finite and infinite regions, a planar graph is often embedded in the surface of sphere. This is done by stereographic projection.

Problems

Let us show that (a) a graph of order 5 and size 8 and (b) a graph of order 6 and size 12, are planar graphs.

Solution:

Given:

A graph of order 5 and size 8 can be represented by a plane drawing.

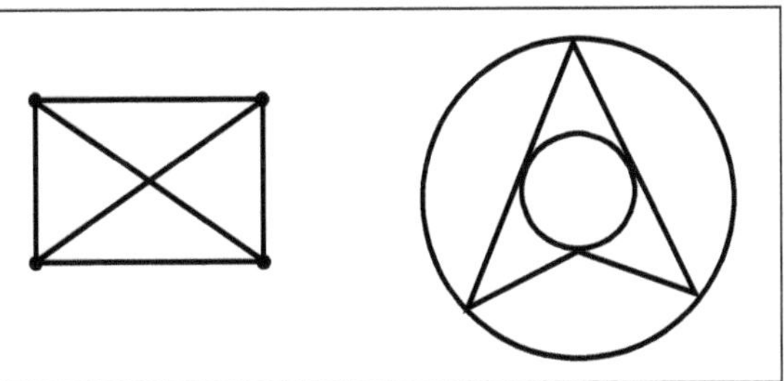

(a) A graph of order 5 and size 8 and (b) a graph of order 6 and size 12.

Here, the edges of the graph meet only at the vertices, as shown in figure (a). Therefore, this graph is a planar graph. Figure (b) shows that a graph of order 6 and size 12 is a planar graph. The plane representations of graphs are by no means unique.

Indeed, a graph G may be drawn in arbitrarily many different ways. The properties of a graph are not necessarily immediate from one representation, but may be apparent from another. However, there are important families of graphs, the surface graphs that rely on the properties of the drawings of graphs.

4.4.1 Graph Coloring

Let us consider a graph G and a positive integer m. If the nodes of G can be colored in such a way that no 2 adjacent nodes have the same color. Yet only 'M' colors are used. So it is called M-color ability decision problem. The graph G can be colored using the least integer 'm'. This integer is referred to as chromatic number of the graph.

A graph is referred to be planar it can be drawn on plane in such a way that no 2 edges cross each other. Suppose we are given a map then, we have to convert it into planar. Consider each and every region as a node. If 2 regions are adjacent then the corresponding nodes are joined by an edge. Consider a map with 5 regions and its graph.

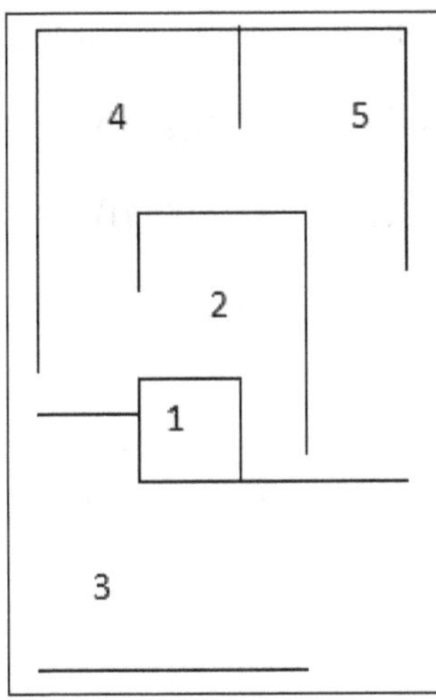

1 is adjacent to 2, 3, 4.

2 is adjacent to 1, 3, 4, 5.

3 is adjacent to 1, 2, 4.

4 is adjacent to 1, 2, 3, 5.

5 is adjacent to 2, 4.

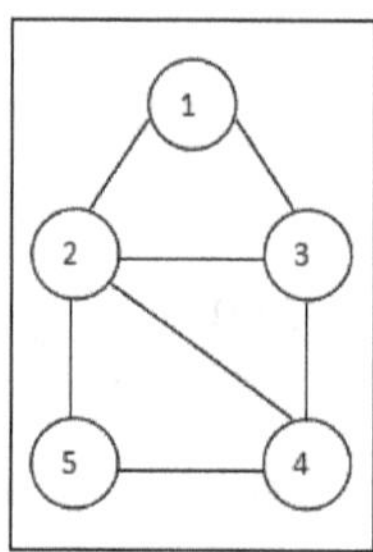

Steps to color the Graph:

- Initially create the adjacency matrix graph(1:m,1:n) for a graph, if there is an edge between i, j then C(i, j) = 1 otherwise C(i, j) =0.
- The Colors will be represented by the integers 1,2,.....m and the solutions will be stored in the array $X_1(1), X_1(2), \ldots\ldots\ldots, X_1(n)$,X(index) is the color, index is the node.
- The formula used to set the color is:
  - $X_1(k) = (X_1(k)+1)\ \%\ (m+1)$
- First one chromatic number is assigned, after assigning a number for 'k' node, we have to check whether the adjacent nodes has got the same values if so then we have to assign the next value.
- Repeat the procedure until all possible combinations of colors are found.
- The function which is used to check the adjacent nodes and same color is:
  - If ((Graph (k, j) == 1) and X (k) = X (j))

Example:

N= 4

M= 3

Adjacency Matrix

0 10 1

1 0 1 0

0101

1010

Problem is to color the stated graph of 4 nodes using three colors. Node-1 can take the given graph of four nodes using 3 colors. The state space tree will give all possible colors in that, the numbers which are inside the circles are nodes and the branch with a number is the colors of the nodes.

State Space Tree

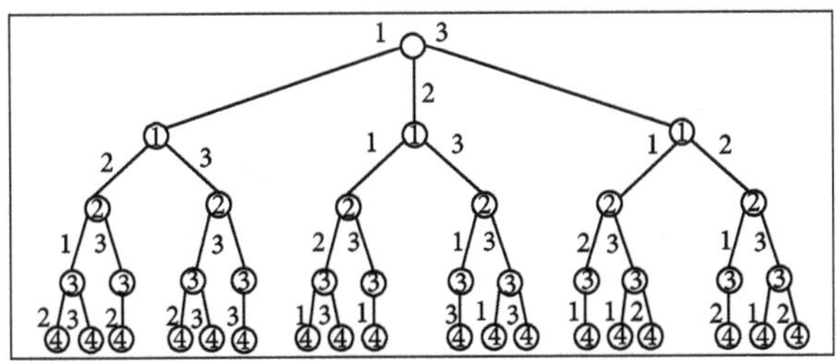

4.5 Trees

Tree is the connected graph without cycles. Graph is called acyclic, if it has no cycles. Acyclic graph is also termed as forest. Tree is a connected acyclic graph.

Theorem

A connected graph is a tree if and only if all its edges are bridges. The graph is a tree. Trees with one, two, three and four vertices are shown in the below figure. A graph must have at least one vertex and therefore so must a tree. Similarly, as we are considering only finite graphs, our trees are also finite.

It follows immediately from the definition that a tree has to be a simple graph, having neither a self-loop nor parallel edge. Trees appear in numerous instances.

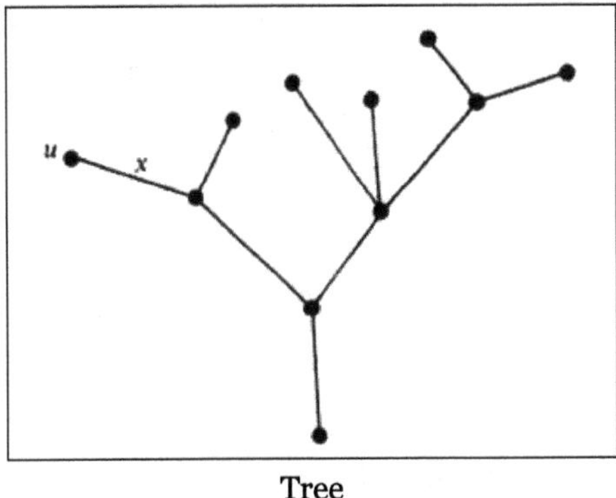

Tree

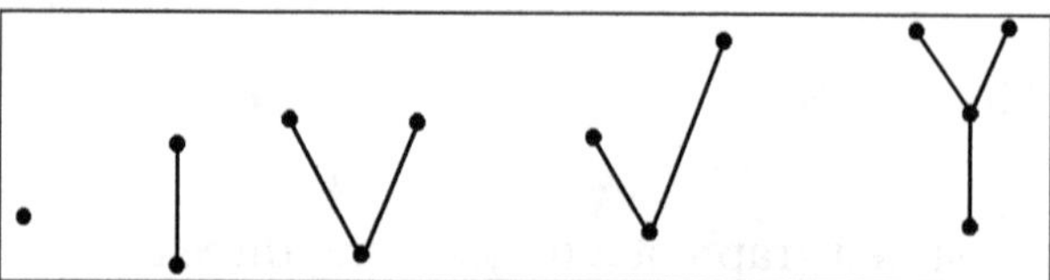

Tree with one, two, three and four vertices.

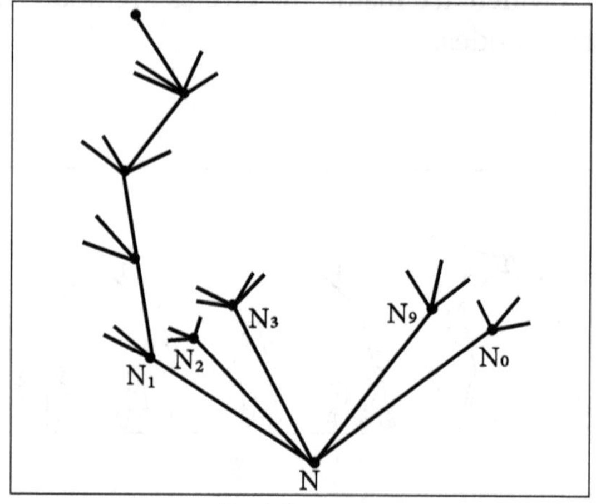

Decision tree.

Basic Properties

A connected graph G is called a tree, if the removal of any of its edges makes G disconnected.

A tree can be defined in a variety of ways as is shown in the following theorem:

Properties of Trees

Theorem 1

There is one and only one path between every pair of vertices in a tree, T.

Proof:

Since T is a connected graph, there must exist at least one path between every pair of vertices in T. Now suppose that between two vertices a and b of T, there may have two distinct paths. The union of these two paths will contain a circuit and T cannot be a tree.

Theorem 2

If in a graph G there is one and only one path between every pair of vertices, G is a tree.

Proof:

Existence of a path between every pair of vertices assures that G is connected. A circuit in a graph (with two or more vertices) implies that there is at least one pair of vertices a, b such that there are two distinct paths between a and b.

Since G has one and only one path between every pair of vertices, G can have no circuit. Therefore, G is a tree.

Distance and Centers in Tree

The below mentioned tree has four vertices. Intuitively, it seems that vertex b is located more "centrally" than any of the other three vertices.

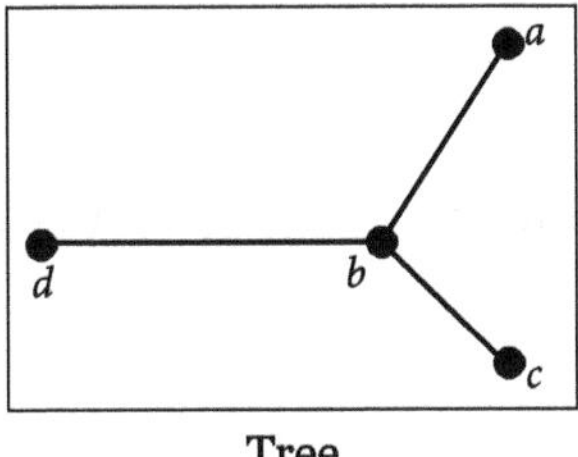

Tree

In a connected graph G, the distance d (v_i v_j) between two of its vertices vi and vj is the length of the shortest path (i.e., the number of edges in the shortest path) between them. The definition of distance between any two vertices is valid for any connected graph (not necessarily a tree). In a graph that is not a tree, there are generally several paths between a pair of vertices.

We have to enumerate all these paths and find the length of the shortest one. (There may be several shortest paths.) For instance, some of the paths between vertices v_1 and v_2 in the figure are (a, e), (a, c, f), (b, c, e), (b, f), (b, g, 1) and (b, g, I, k). There are two shortest paths, (a, e) and (b, f), each of length two. Hence, d (v_1, v_2) = 2.

In a tree, since there is exactly one path between any two vertices, the determination of distance is much easier. For instance, in the tree, d (a, b) = 1, d (a, c) = 2, d (c, b) = 1 and so on.

A Metric

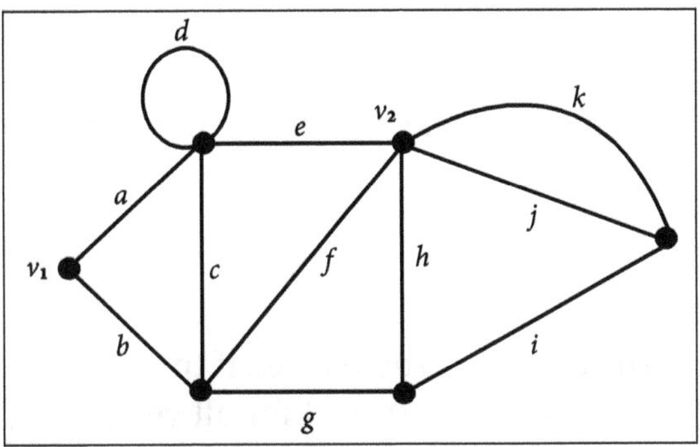

Distance between v_1 and v_2 in 2.

Before we can legitimately call a function f(x, y) of two variables as a "distance" between them, this function must satisfy certain requirements.

These are:

- Non-negative: $f(x, y) \geq 0$, and $f(x, y) = 0$ if and only if $x=y$.
- Symmetry: $f(x, y) = f(y, x)$.
- Triangle inequality: $f(x, y) \leq f(x,z) \leq f(x, z) + f(z, y)$ for any z.

Theorem

The distance between vertices of a connected graph is a metric.

Eccentricity

The eccentricity E (v) of a vertex v in a graph G is the distance from v to the vertex farthest from v in G. i.e.,

$$E(v) = \max_{vi \in G} d(v, v_1)$$

A vertex with minimum eccentricity in graph G is called a center of G. The eccentricities of the four vertices are E (a) = 2, E (b) = 1, E (c) = 2 and E (d) 2. Hence, vertex b is the center of that tree.

On the other hand, consider the tree in Figure. The eccentricity of each of its six vertices is shown next to the vertex. This tree has two vertices having the same minimum eccentricity. Hence, this tree has two centers. Some refer to such centers as bicenters. We can easily verify that a graph, in general, has many centers. For example, in a graph that consists of just a circuit (a polygon), every vertex is a center.

4.5.1 Rooted Trees

Directed Tree

If G is a directed graph, then G is called a directed tree if the undirected graph associated with G is a tree.

Rooted Tree

If G is a directed tree, then G is called rooted tree if there is unique vertex r, called the root, in G with in-degree of r = id(r) = 0, and for all vertices, v, then in-degree of v = id(v) = 1.

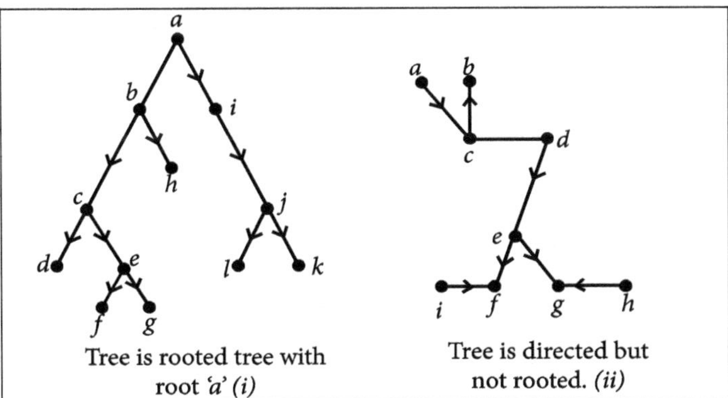

Tree is rooted tree with root 'a' (i)

Tree is directed but not rooted. (ii)

A rooted tree is a tree with a designated vertex called the root of the tree.

- In the rooted tree, the directions are understood as going from upper level to the lower level so that arrow are not needed.
- In a rooted tree, a vertex v without degree od (v) = 0 is called leaf or terminal vertex.

A rooted tree is a tree in which a particular vertex is distinguished from the others and that particular vertex is called the root. In graph theory, we consider the edge of a tree pointing downward and rooted trees are typically drawn with their root at the top.

To form a rooted tree, first we place the root at the top. Under the root and on the same level, we place the vertices that can be reached from the root on the same path of length 1. Under each of these vertices and on the same level, we place vertices that can be reached from the root on a simple path of length 2. We continue this process until the entire tree is drawn.

Ordered Rooted Tree

An ordered rooted tree is a rooted tree whose sub trees are put into a definite order and are known as ordered trees. An empty and a single vertex are ordered rooted trees.

Level of a Vertex

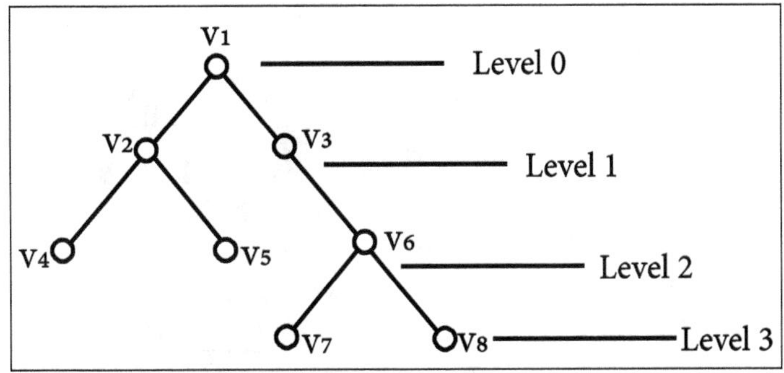

The level of a vertex is the number of edges along the unique path between it and the root. The level of the root is defined as zero. The vertices immediately under the root are said to be in level 1 and so on.

In the figure:

V_1 is root.

V_2 and V_3 are at level one.

V_4, V_5 and V_6 are at level two.

V_7, V_8 are at level three.

Height

The height of a rooted tree is the maximum level to any vertex of the tree. The depth of a vertex V in a tree is the length of the path from the root V. For example, the height of the above tree is 3 and depth of V_6 is 2.

Children, Parent and Siblings

Given any internal vertex V of a rooted tree, the children of V are all those vertices that are adjacent to V and are one level farther away from the root than V. If W is a child of V, then V is called the parent of W, and two vertices that are children of the same parent are called the siblings.

If the vertex u has no children, then u is called a leaf or a terminal node (vertex). If u has either one or two children, then u is called an internal vertex.

Descendants and Ancestor

The descendants of the vertex V_i is the set consisting of all the children of V_i together with the descendants of those children. Given vertices V_j and V_k, if V_j lies on the unique path between V_k and the root, then V_j is an ancestor of V_k and V_k is a descendant of V_j.

Example:

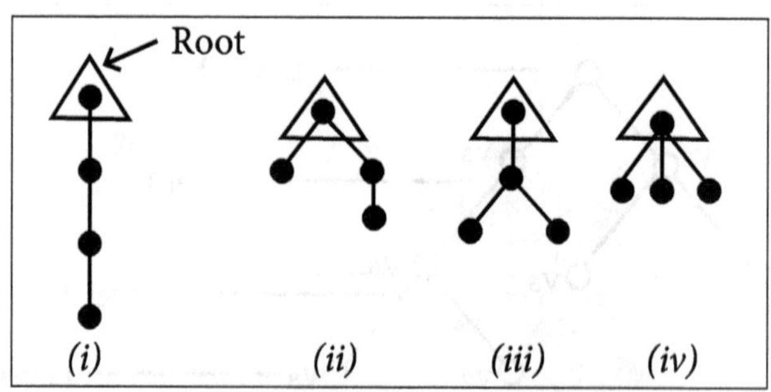

Rooted Trees.

Descendants and Ancestors

In the graph as shown in figure (i), the vertices e, f, g are called descendants of c, b, a and c, b, a are called ancestors of e, f and g.

Sibling

The vertices with a common parent are referred as siblings. In figure (i) f and g are siblings of b.

4.5.2 Binary Search Trees

Binary & m-ary Trees

We describe a directed tree as binary if no vertex has out degree over 2. It is more common to call a tree binary if no vertex has degree over 3. In general, a tree is m-ary if no vertex has degree over m+1. We call a directed tree as m-ary if no vertex has out degree over m. In a rooted binary tree (hanging down or growing up), one can describe each child vertex as the left child or right child of its parent.

Binary Search Tree

A binary search tree is a binary tree in which each child of a vertex designated as a right or left child, no vertex has more than one right child or left child, and each vertex is labeled with a key, which is one of the items.

Vertices are assigned keys so that the key of a vertex is both larger than the keys of all vertices in its left sub tree and smaller than the keys of all vertices in its right sub tree.

Decision Trees

A rooted tree in which each internal vertex corresponds to a decision with a sub tree at these vertices for each possible outcome of the decision is called a decision tree.

Trees Traversal

We will describe three of the most commonly used algorithms such as preorder traversal, in order traversal and post order traversal.

Let T be an ordered rooted tree with root r. If T consists only of r, then r is the preorder traversal of T. Otherwise, suppose that T_1, T_2,...T_n are the sub trees at r from left to right in T. The preorder traversal begins by visiting r. It continues by traversing T_1 in preorder, then T_2 in preorder and so on until T_n is traversed in preorder. Let T be an ordered rooted tree with root r. If T consists only of r, then r is the in order traversal of T.

Otherwise, suppose that T_1, T_2, . . . , T_n, are the sub trees at r from left to right. The in order traversal begins by traversing T_1 in in order, then visiting r. It continues by traversing T_2 in in order, then T_3 in in order, . . . , and finally T_n in in order.

Let T be an ordered rooted tree with root r. If T consists only of r, then r is the post order traversal of T. Otherwise, suppose that T_1, T_2, . . . , T_n are the sub trees at r from left to right.

The post order traversal begins by traversing T_1 in post order, then T_2 in post order, ..., then T_n in post order, and ends by visiting r.

4.5.3 Spanning Trees

If G (V, E) is a graph and T (V, F) is a sub graph of G and is a tree, then T is a spanning tree of G. That is, T is a tree that includes every vertex of G and has only edges to be found in G. Using the procedure in the directed graph, we can easily prove that every connected graph has a spanning tree.

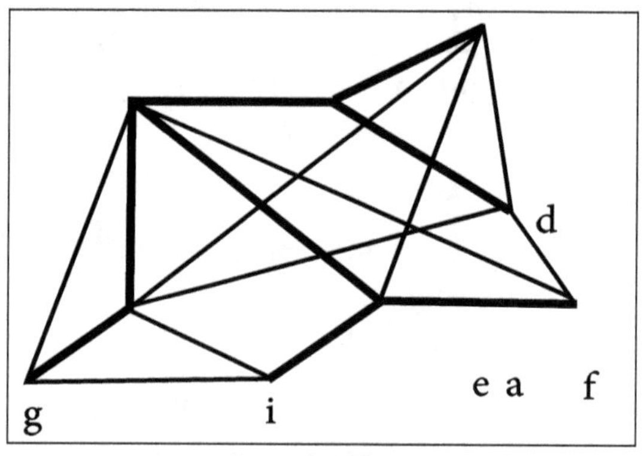

Spanning Tree.

Every connected graph has a spanning tree which can be obtained by removing edges until the resulting graph becomes acyclic. In practice, however, removing edges is not efficient because finding cycles is time consuming.

Next, we give two algorithms to find the spanning tree T of a loop-free connected undirected graph G — (V, E). We assume that the vertices of G are given in a certain order v_1, v_2, ... v_m. The resulting spanning tree will be T — (V_i, E_i).

Spanning Trees

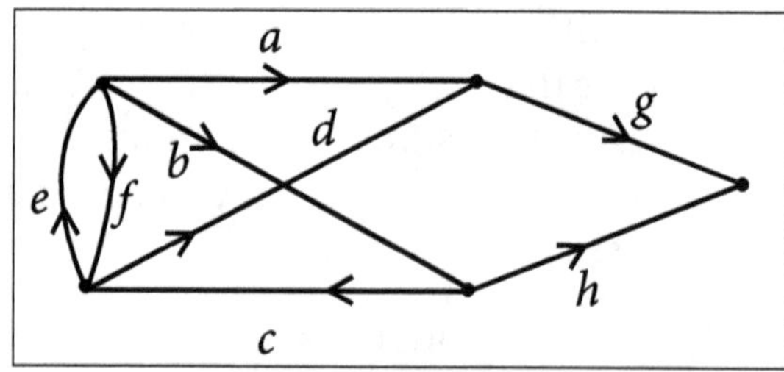

A spanning tree is an n-vertex connected digraph analogous to a spanning tree in an undirected graph and consists of n – 1 directed arcs. A spanning arborescence in a connected digraph is a spanning tree that is an arbore scene. For example, {a, b, c, g} is a spanning arborescence in the figure.

Theorem

In a connected isograph D of n vertices and m arcs, let W = (a_1, a_2,..., a_m) be an Euler line, which starts and ends at a vertex v (that is, v is the initial vertex of a1 and the terminal vertex of a_m). Among the m arcs in W there are n – 1 arcs that enter each of n−1 vertices, other than v, for the first time. The sub digraph D_1 of these n−1 arcs together with the n vertices is a spanning arborescence of D, rooted at vertex v.

Proof:

In the sub digraph D_1, vertex v is of in degree zero and every other vertex is of in degree one, for D_1 includes exactly one arc going to each of the n−1 vertices and no arc going to v. Further, the way D_1 is defined in W, implies that D_1 is connected and contains n−1 arcs. Therefore D_1 is a spanning arborescence in D and is rooted at v.

Illustration

In Figure, W = (b d c e f g h a) is an Euler line, starting and ending at vertex 2. The sub digraph {b, d, f} is a spanning arborescence rooted at vertex 2.

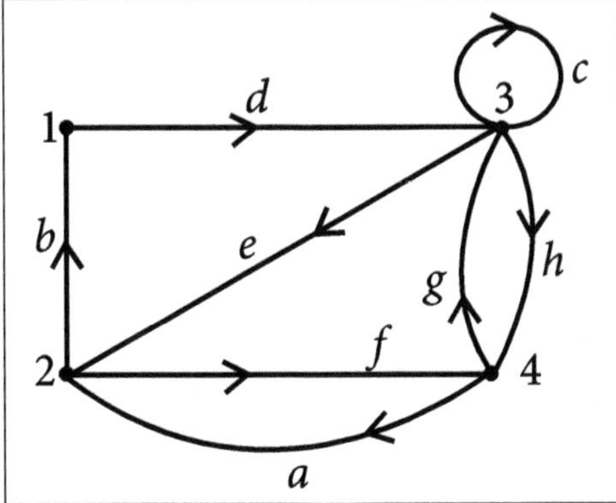

4.5.4 Minimum Spanning Trees

A Spanning tree of an undirected graph, G is a tree formed from graph edges that connects all vertices of G. A Minimum Spanning tree of an undirected graph, G is a tree formed from graph edges that connects all vertices of G at lowest cost.

A minimum spanning tree exists if and only if G is connected. The number of edges in the minimum spanning tree is |V| - 1. The minimum spanning tree is a tree because it is acyclic, it is spanning because it covers every vertex and it is minimum because it covers with minimum cost. The minimum spanning tree can be created using two algorithms, that is prim's algorithm and kruskal's algorithm.

Suppose G = (V, E) is a graph whose edges are assigned lengths (or weights) that is, are labelled with positive numbers. The minimal spanning tree T of G is that spanning tree that has the smallest length sum (weight sum) among all the spanning trees.

Definition

A minimal spanning tree of G is a spanning tree of G with minimum weight. There are several methods (algorithm) available for actually finding the shortest spanning tree in a given graph, both by hand and computer.

4.5.4 Kruskal's Algorithm

Strategy

- The edges are build into a min heap structure and each vertex is considered as a single node tree.
- The Delete Min operation is utilized to find the minimum cost edge (u, v).
- The vertices u and v are searched in the spanning tree set S and if the returned sets are not same then (u, v) is added to the set S (union operation is performed), with the constraint that adding (u, v) will not create a cycle in spanning tree set S.
- Repeat step (ii) & (iii), until a spanning tree is constructed with |v|-1 edges.

Example:

Given:

G = (V, E, W)

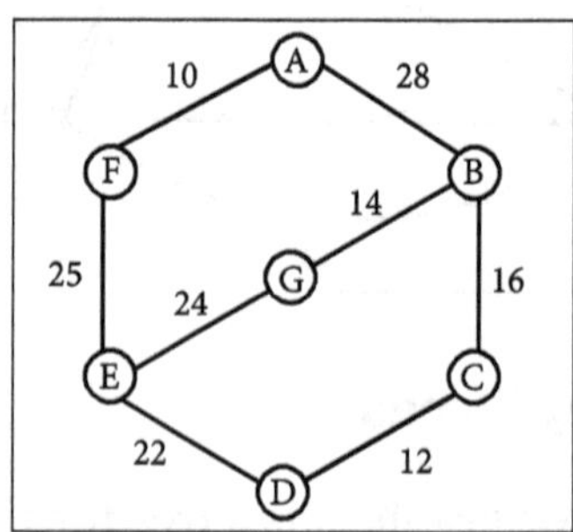

Edge	Weight	Action	Comments
(A, F)	10	Accepted	An edge with minimum cost. (from Delete Min of heap)
(C, D)	12	Accepted	
(B, G)	14	Accepted	
(B, C)	16	Accepted	

(D, E)	22	Accepted	
(E, G)	24	Rejected	Forms a cycle
(F,E)	25	Accepted	
(A, B)	28	Rejected	Forms a cycle

Action of Kruskal's algorithm on G.

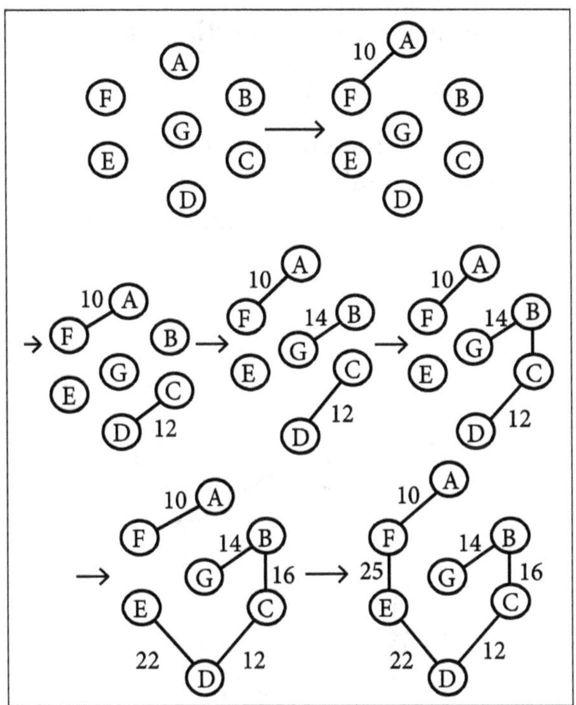

Kruskal's algorithm after each stage.

At each stage, an edge with minimum cost is selected by performing Delete Min from heap structure, if it doesn't form a cycle.

Analysis

The worst-case running time of this algorithm is O (|E| log |E|), with heap operations, since |E| = o (|V|²), this running time is actually O (|E| log |V|).

4.5.5 Prim's Algorithm

In this method, minimum spanning tree is constructed in successive stages. One node is picked as the root and an edge is added (i.e.,) an associated vertex is added to the tree, until all the vertices are present in the tree with |V|-1 edges.

The Strategy

- One node is picked as a root node (u) from the given connected graph.

- At each stage choose a new vertex v from u, by considering an edge (u, v) with minimum cost among all edges from u, where u is already in the tree and v is not in the tree.
- The Prim's algorithm table is constructed with three parameters. They are:
  - known- Vertex is added in the tree or not.
  - dv - Weight of the shortest arc connecting v to a known vertex.
  - pv- last vertex which causes a change in dv.
- After selecting the vertex v, the update rule is applied for each unknown w adjacent to v. The rule is $d_w = \min(d_w, C_{w,v})$, that is if more than one path exist between v to w, then d_w is updated with minimum cost.

Example:

Given:

G = (V, E, W)

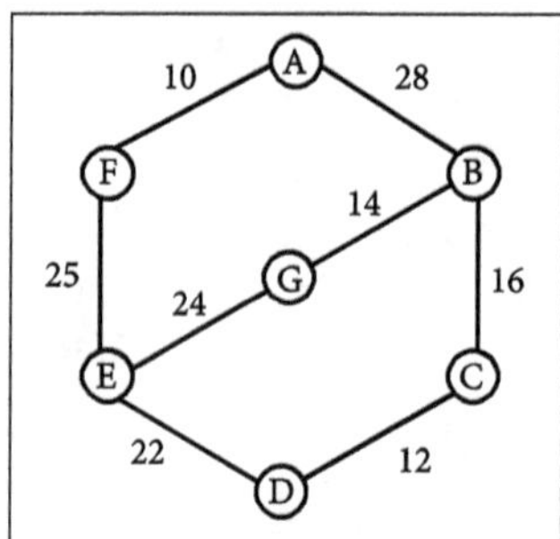

The sequence of edges added at each stage from the start vertex A is given below:

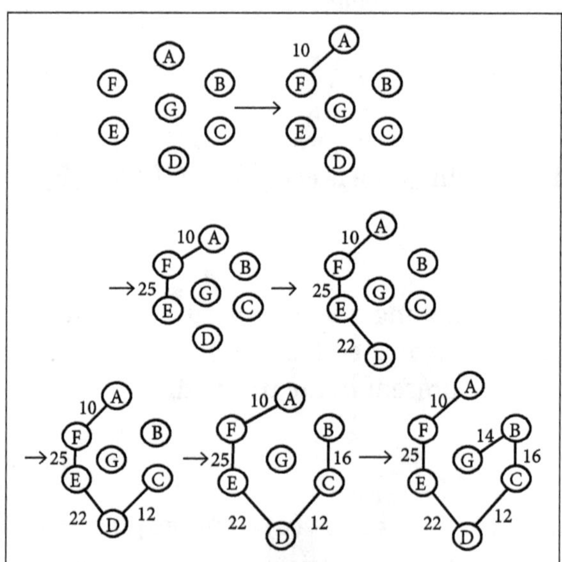

The above discussed example is explained using a table with three parameters. They are known, dv and pv.

V	known	dv	pv
A	0	0	0
B	0	∝	0
C	0	∝	0
D	0	∝	0
E	0	∝	0
F	0	∝	0
G	0	∝	0

Initial configuration of table for the given example using Prim's algorithm.

V	known	dv	pv
A	1	0	0
B	0	28	A
C	0	∝	0
D	0	∝	0
E	0	∝	0
F	0	10	A
G	0	∝	0

The table after A is declared known.

∴ Known (A) = 1, other vertices remains zero.

dv (B) = 28, cost from A to B.

pv (B) = A, the vertex to reach B.

The table after F is declared known.

V	known	dv	pv
A	1	0	0
B	0	28	A
C	0	∝	0
D	0	22	E
E	1	25	F
F	1	10	A
G	0	24	E

The table after E is declared known:

V	known	dv	pv
A	1	0	0
B	0	28	A
C	0	12	D
D	1	22	E
E	1	25	F
F	1	10	A
G	0	24	E

The table after D is declared known:

V	known	dv	pv
A	1	0	0
B	0	16	C
C	1	12	D
D	1	22	E
E	1	25	F
F	1	10	A
G	0	24	E

The table after C is declared known. Here the dv value of B is updated using update rule. That is B can be reached through either A (or) C, from this the edge with minimum cost is chosen (i.e.) CB = 16.

The table after B is declared known:

V	known	dv	pv
A	1	0	0
B	1	16	C
C	1	12	D
D	1	22	E
E	1	25	F
F	1	10	A
G	0	14	B

In the above table, dv value of G is updated using update rule.

V	known	dv	pv
A	1	0	0
B	1	16	C
C	1	12	D

D	1	22	E
E	1	25	F
F	1	10	A
G	1	14	B

The table after G is declared known.

(Prim's algorithm terminates with | V|-1 edges).

The edges in the spanning tree can be read from the table: (A, F) (F, E) (E, D) (D, C) (C, B) (B, G). The total number of edges is 6 that is equivalent to |V|-1 (No. of vertices-1).

Permissions

All chapters in this book are published with permission under the Creative Commons Attribution Share Alike License or equivalent. Every chapter published in this book has been scrutinized by our experts. Their significance has been extensively debated. The topics covered herein carry significant information for a comprehensive understanding. They may even be implemented as practical applications or may be referred to as a beginning point for further studies.

We would like to thank the editorial team for lending their expertise to make the book truly unique. They have played a crucial role in the development of this book. Without their invaluable contributions this book wouldn't have been possible. They have made vital efforts to compile up to date information on the varied aspects of this subject to make this book a valuable addition to the collection of many professionals and students.

This book was conceptualized with the vision of imparting up-to-date and integrated information in this field. To ensure the same, a matchless editorial board was set up. Every individual on the board went through rigorous rounds of assessment to prove their worth. After which they invested a large part of their time researching and compiling the most relevant data for our readers.

The editorial board has been involved in producing this book since its inception. They have spent rigorous hours researching and exploring the diverse topics which have resulted in the successful publishing of this book. They have passed on their knowledge of decades through this book. To expedite this challenging task, the publisher supported the team at every step. A small team of assistant editors was also appointed to further simplify the editing procedure and attain best results for the readers.

Apart from the editorial board, the designing team has also invested a significant amount of their time in understanding the subject and creating the most relevant covers. They scrutinized every image to scout for the most suitable representation of the subject and create an appropriate cover for the book.

The publishing team has been an ardent support to the editorial, designing and production team. Their endless efforts to recruit the best for this project, has resulted in the accomplishment of this book. They are a veteran in the field of academics and their pool of knowledge is as vast as their experience in printing. Their expertise and guidance has proved useful at every step. Their uncompromising quality standards have made this book an exceptional effort. Their encouragement from time to time has been an inspiration for everyone.

The publisher and the editorial board hope that this book will prove to be a valuable piece of knowledge for students, practitioners and scholars across the globe.

Index

Printed in the USA
CPSIA information can be obtained
at www.ICGtesting.com
JSHW051349091023
49903JS00006B/78